The Nature of

ZIMBABWE

A GUIDE TO CONSERVATION AND DEVELOPMENT

The Nature of Zimbabwe is the third volume in a series of information books about conservation and development issues around the world. Called The Nature of . . . series, it is not a profit-making venture, but forms part of the education and awareness work of IUCN's Field Operations Division. This edition was produced in partnership with the IUCN Regional Office for Southern Africa.

The Nature of Zimbabwe was made possible by the generosity of Grindlays Bank in Zimbabwe, and by the Royal Norwegian Ministry of Development Cooperation.

Field Operations Division
IUCN
Avenue du Mont-Blanc
1196 Gland
Switzerland

IUCN Regional Office for Southern Africa
P.O. Box 745
Harare
Zimbabwe

ISBN 2-88032-933-7

Compiled by David Jones

Edited and produced by Mark Carwardine

Design and illustrations © Christine Bass

Typeset by SP Typesetting, Birmingham, England and printed by Albany House Limited, Coleshill, England.

All the photographs were taken by David Reed (DR), with the following exceptions noted in the text:

Stephen Bass (SB): 17
Ilo Battigelli (IB): 73
Hans Christen (HC): 29, 72
Meg Coates Palgrave (MCP): 42, 43
David Hartung (DH): 17, 26, 30, 42, 43
Michael Kimberley (MK): 30
Photobank (PB): 24, 53
Dick Pitman (DP): 31
Zimbabwe Tourist Development Corporation (ZTDC): 37

Further copies of titles in *The Nature of . . .* series are available from the Field Operations Division at the above address.
Price: US$7.50/£5.00/Sfr15.00 (or equivalent) plus post and packing.

Other titles in this series:
The Nature of Pakistan
The Nature of Zambia

In preparation:
The Nature of Bangladesh
The Nature of Botswana
The Nature of Costa Rica
The Nature of Kenya
The Nature of Panama

Founded in 1948, the International Union for Conservation of Nature and Natural Resources (IUCN) is the world's largest and most experienced alliance of active conservation authorities, agencies and interest groups. Its 634 members include States, government departments and most of the world's leading independent conservation organisations, in over 120 countries.

The Field Operations Division (formerly the Conservation for Development Centre – CDC) is the entrepreneurial arm of IUCN. It specialises in the application of conservation principles to the process of economic development. Since CDC's establishment, in 1981, it has become known worldwide for its work in ensuring that the exploitation of natural resources is done on a sustainable basis – to ultimately yield greater benefits for mankind – rather than for short-term gain. A small staff of under twenty, based in Gland, Switzerland, coordinates a worldwide network of consultants and experts for this work and is currently establishing regional centres in a number of developing countries.

Up to five million litres of River Zambezi water cascade every second over the one and a half kilometre width of Victoria Falls. The roar of the waters and the great clouds of spray (Mosi-Oa-Tunya or "The smoke that thunders") make the visitor keenly aware of the great power of nature.

Foreword

The appearance of *The Nature of Zimbabwe* is most welcome. It celebrates the great natural beauty and diversity of our country; it records the achievements of our people; and it seeks to open eyes and minds to the dangers of environmental damage that grow daily around us. Above all, it emphasises the need for conservation.

Never has the need for conservation awareness been more urgent. The land on which we all depend for survival and development is losing its own vitality under the increasing burdens we place upon it. The warning signs are becoming more frequent – in the bare and eroded soil, the destroyed trees, the silted rivers and dams, the poor crops. These signs point to impending social and economic problems. They must be heeded now to ensure sustainable development for present and future generations.

Our fast growing population must feed, clothe and house itself, and our people must be cared for medically, educated, and employed. They must live in the full sense, not just struggle to survive. Our country, although impressively productive, is very vulnerable. With its predominantly poor soils and low rainfall, drought is never far from our shoulder.

Our Government is deeply committed to the development of the rural areas, where most of the people live and where the need is greatest. But development is impossible if the land cannot support it. 'Natural resource management for sustainable development' is the conservation concept that is now sweeping the world. Zimbabwe subscribes to this concept but is aware of the difficulties of implementing it due to the history of the country which severely restricted the majority of the inhabitants' access to resources. This is coupled with unequal terms of trade and destabilisation policies exercised by the apartheid regime of South Africa. Conservation for sustainable development simply means that the things provided by nature must be looked after and used well, because they are the foundation of life and therefore of development. In certain situations, where development and conservation inevitably clash head-on, the challenge is to reconcile the two, recognising their dependence on each other.

Precariously balanced climatically and environmentally though it is, Zimbabwe is also rich in resources and is better placed than many developing countries to get on top of its problems. The conservation ethic is traditional for us; much has been done by dedicated people over many years, with good results. But the great need now, as rapidly increasing population pressures threaten to tip the scale against us, is for a truly national effort on the part of all Zimbabweans to safeguard their land and their future.

Our own National Conservation Strategy shows the road to survival. We have the resources and the ability to set out on that difficult road – whether we reach our destination will depend on the will and motivation of Zimbabweans themselves. That motivation begins with people in positions of responsibility and influence in many walks of life – the readers of *The Nature of Zimbabwe*.

I thank IUCN which, through the Regional Office, has initiated this publication; the sponsors within and outside Zimbabwe for making it possible; and all who contributed to its excellence.

Hon. Victoria Chitepo
Minister of Natural Resources and Tourism

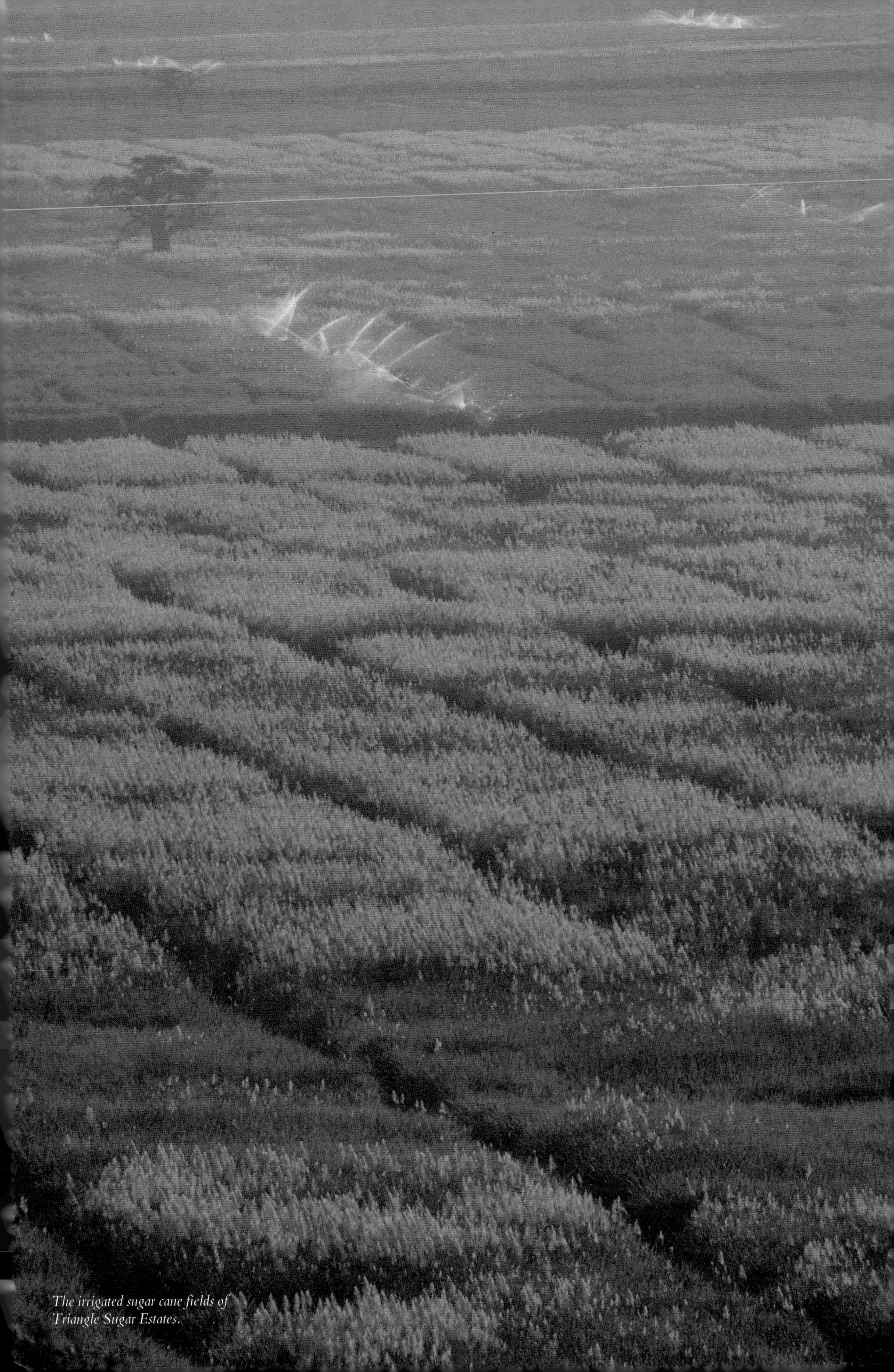

The irrigated sugar cane fields of Triangle Sugar Estates.

Contents

Chapter One
ZIMBABWE: ITS LAND AND PEOPLE

History

As long as 40,000 years ago there were people chipping away at stones in the hills and valleys that are now part of Zimbabwe. Thousands of Stone Age archaeological sites disclose some of their secrets. Distinctive rock art, with its elegant and elongated figures on the granite walls of caves and rock shelters, is further dramatic evidence of other early inhabitants of the region – the diminutive hunter-gatherers who arrived here several thousand years later.

But Zimbabwe's history has been a turbulent one – and few of the people living in the country today can claim descent from its earliest inhabitants. Most are heirs to the pioneers and conquerors, colonisers and immigrants – both black and white – who have passed through or settled here over the centuries.

The Shona people have the longest history in Zimbabwe. They are thought to be descendants of second century AD Iron Age inhabitants, who migrated from the north onto the Zimbabwean plateau, between the Zambezi and the Limpopo rivers. Shona settlement of the plateau was established by about 900 in the south and 1100 in the north.

By the eleventh century fairly elaborate forms of economic and social organisation had developed in the Shona communities, leading to an era in which the Zimbabwean tradition of building in stone flourished. The principal achievement of this period was the construction (between the thirteenth and fifteenth centuries) of what is now known as Great Zimbabwe. This impressive array of stone-walled enclosures was home to as many as 20,000 people – and, at the time, was the largest city in sub-Saharan Africa. A feature of Shona history has been the formation of states; Great Zimbabwe was the capital of the state known as Zimbabwe, its wealth based on the domination of the trade routes between the goldfields and the east coast.

But by the end of the fifteenth century, the soil around Great Zimbabwe had become impoverished, the nearest timber was too far away and many of the surface gold workings on the plateau had been worked out. Many of the people moved away and the stone capital shrank to regional importance in the new Kingdom of Munhumutapa.

Then, in 1510, Portuguese traders began to visit the area. Armed conflict arose as they strengthened their position, and this led ultimately to the decline of the Munhumutapa Kingdom in the early 1800s. At about the same time, there was a series of invasions from the south by the Ngoni people, an offshoot of the powerful Zulu warrior tribe. They established the Ndebele state, which absorbed many smaller tribes.

Missionaries – David Livingstone among them – were sporadically active in the region for much of the nineteenth century. The first permanent white settlement, the Inyati Mission, was established in 1859. But the build-up of white settlers did not really gather pace until some 30 years later, in the wake of the discovery of gold in Mashonaland in 1867.

It was in 1889 that the die was cast for the land that was to become known as Rhodesia. In that year Cecil John Rhodes, pursuing his dream of extending British rule from the Cape to Cairo, obtained a Royal Charter over the territory for his British South Africa Company (BSAC). The fol-

Opposite:
Eighty per cent of Zimbabweans live in rural areas, depending closely on the natural environment for food, fuel, timber and water.

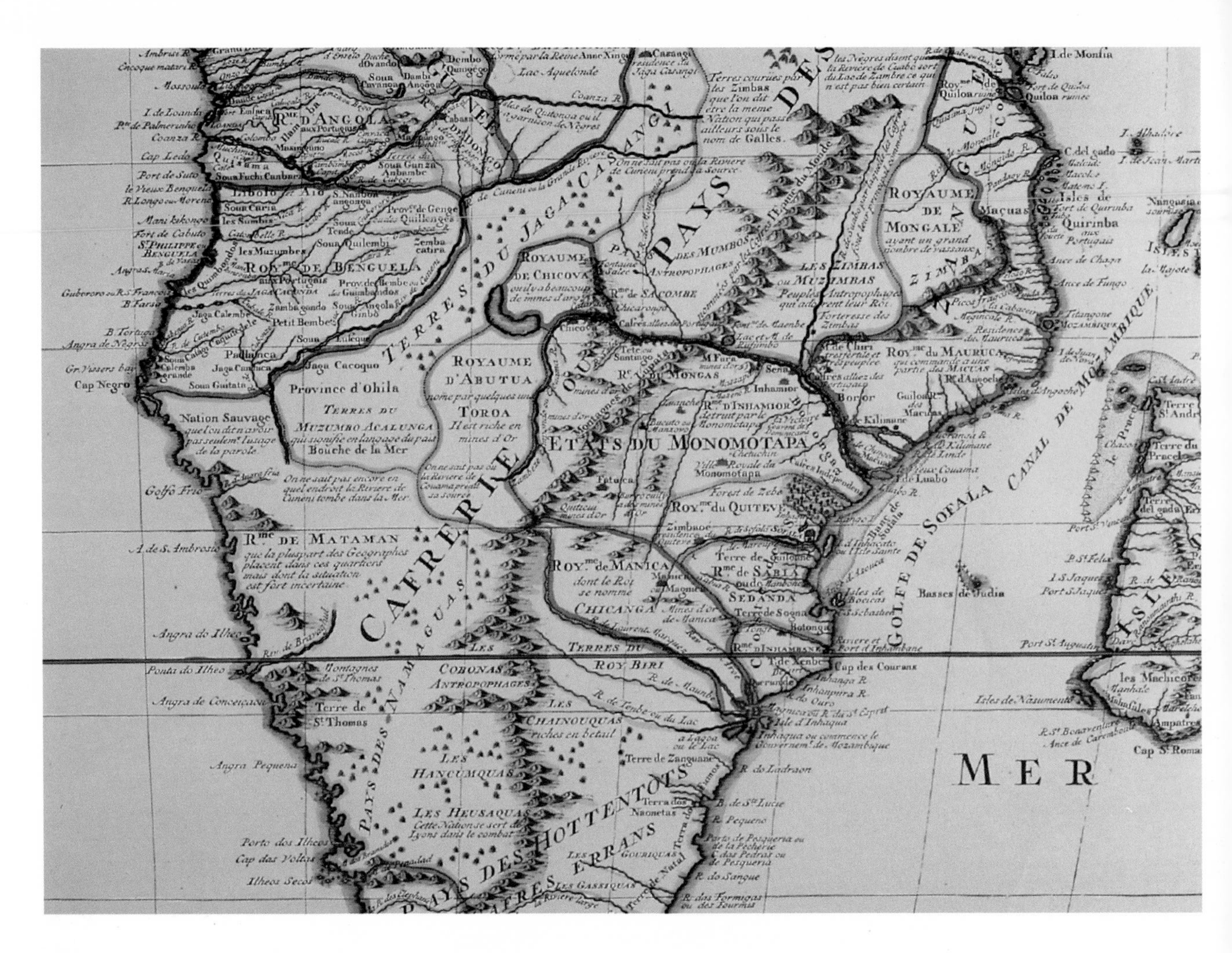

This 1720 map by G. de l'Isle shows a rudimentary European knowledge of the Monomotapa Kingdom, which occupied much of present-day Zimbabwe. (Collection of the National Archives of Zimbabwe).

lowing year he became Prime Minister of the Cape and the 'Pioneer Column' of the BSAC – a heavily armed force consisting of a corps of 212 men, five troops of the BSA Company Police, 200 black troops from Bechuanaland (Botswana) and 16 civilians (mainly gold prospectors) – invaded and occupied Mashonaland. On 13 September 1890 the Union Jack was raised at Fort Salisbury, in the centre of what is now the capital, Harare.

Land acquisition started immediately. The occupation of Matabeleland took place three years later, overcoming strong but futile resistance from the Ndebele. In 1896 the Ndebele rose again and no sooner had this uprising been quelled than a second, more serious, revolt broke out in Mashonaland. This too was crushed, and its leaders executed. Zimbabweans today call these uprisings the First Chimurenga (which in Shona means the first 'war of liberation') and view them as the inspiration for the Second Chimurenga – the guerrilla war which was conducted to gain independence nearly 90 years later.

On 12 September 1923 Southern Rhodesia (as it was known) was transferred from the BSAC to the British Empire and became a colony. The Land Apportionment Act of 1930 and the Industrial Conciliation Act of 1934 ensured that the main role of the Africans in the economy was as labour migrants to the towns, farms and mines. The legislation was amended many times but remained constant in its application, and was the system which the present Government inherited at independence in 1980. This early division of the country, for occupation on racial lines, was the root cause of much of the land damage that has occurred in Zimbabwe in the past 60 years.

In 1953 Southern Rhodesia, Northern Rhodesia (Zambia) and Nyasaland (Malawi) were amalgamated into the much-opposed Federation of Rhodesia and Nyasaland. The Federation was dissolved ten years later, paving the way for the independence of Malawi and Zambia in 1964.

But the British government refused to decolonise Southern Rhodesia until some accommodation had been worked out between the blacks and whites. Ian Smith became Prime Minister on 13 April 1964 and, with overwhelming support from the whites and following unsuccessful negotiations with the British government, made a unilateral declaration of

Above left:
The granite shield that covers much of Zimbabwe provided natural fortress sites and building stone for early civilisations.

Above:
Zimbabwe's ancient rock art is world-renowned. Cave paintings – such as here at Markwe near Marondera – depict man in his natural environment.

independence (UDI) on 11 November 1965.

Britain and the United Nations responded with economic sanctions, and the groundswell of African nationalism – which had been active throughout – developed into full-scale civil war. The main nationalist organisations involved were the Zimbabwe African National Union (ZANU) and the Zimbabwe African People's Union (ZAPU).

There were several abortive attempts to resolve the situation. In 1974, for example, some of the imprisoned nationalists, including Nkomo, Sithole and Mugabe, were released. Five years later, he persuaded some nationalist leaders to cooperate in elections, under a constitution which would guarantee effective white dominance; Muzorewa became Prime Minister, with Ian Smith included in the Cabinet. But the new government failed to win international recognition and the guerrilla war intensified, now with Mugabe's ZANU and Nkomo's ZAPU working as a three-year-old alliance called the Patriotic Front.

In late 1979, following an intense Commonwealth summit meeting, the Patriotic Front, the Salisbury government and the British government agreed on a new constitution for an independent republic of Zimbabwe. There would be a cease-fire to end the war, a three-month period of direct British rule and then elections for a 100-seat House of Assembly, with 20 seats being reserved for whites.

The two parties in the Patriotic Front contested the elections separately. ZANU (PF) won an overall majority and Robert Mugabe was appointed Prime Minister. Joshua Nkomo, whose PF (ZAPU) party won 20 seats, accepted the invitation to join a coalition government. On 18 April 1980, Zimbabwe became legally independent.

The new Prime Minister's strong advocacy of national unity based on reconciliation between former belligerents, and on the repudiation of racial, tribal and social privilege, was a powerful influence in the critical post-independence period. In January 1988 he became executive President of Zimbabwe, which now has a one-party participatory democracy form of government following the merger of ZANU (PF) and (PF) ZAPU.

The Great Zimbabwe Ruins

For Zimbabweans, a visit to the stone ruins of Great Zimbabwe, near the town of Masvingo in the south-east, is something of a pilgrimage. No place, object, image or emblem symbolises the modern state more powerfully and emotively than this ancient city from which Zimbabwe derives its name.

In these great walls, towers, passageways and enclosures, hewn and set in place on an undulating valley floor six centuries ago, black Zimbabweans look back directly on their heritage.

The images of Great Zimbabwe are everywhere. One of its most famous sculptures, a slender stone bird, is the national badge and appears on the flag and on coins and medals. The central Conical Tower in the Great Enclosure is seen on stamps and banknotes and is used as a trademark. And during the independence war in the 1970s, these enduring stone walls were the psychological rallying-point for refugees and Zimbabweans in exile.

It is a still and brooding place. Bees work in the gnarled old red milkwood trees, inside the 250-metre long wall that encircles the Great Enclosure. Small lizards dart along the regimented stones streaked with lichen, tufted with moss and laced with ferns. Butterflies fly silently between the weathered ruins.

This great circle of stones, each faced and butting onto the next with the precision of a brick wall, was once the centre of the largest city in sub-Saharan Africa. In the sprawling complex of surrounding stone structures, extending over more than 700 hectares, lived a well-ordered community of, at any one time, up to 20,000 people. Among them were farmers, herders, soldiers, craftsmen, builders, miners, traders and administrators.

Great Zimbabwe, or *Dzimbabwe* as it is known traditionally (in Shona it means 'houses of stone') is impressive by any standards. The skill of the masons is striking; and the style of the free-flowing architecture is unique to Zimbabwe. Bringing men together to work on such a scale was an impressive achievement.

It is not a fortress but a palace, the grandest of nearly a hundred lesser stone *mazimbabwe* built around the country – a prime example is Khami, near Bulawayo – when Great Zimbabwe was in its prime. A handful of these smaller centres, which follow much the same style of construction, were themselves capitals of independent states.

Archaeologists have progressively stripped away the once enigmatic face of Great Zimbabwe, ending a period of scholarly dispute over its origins. The city was built by local Shona-speaking people and is indigenous in every sense to Zimbabwe.

What remains in dispute is why, after a long period of growth and prosperity, it faltered and declined. Some time late in the fifteenth century Great Zimbabwe lost its political importance. Trade declined, building stopped and the population dispersed. Some experts point out that a community of that size would have put severe strains on the local environment – for food, fuel and building materials. But Great Zimbabwe had managed to avoid these pitfalls for centuries. Historians suggest that the answer might lie in a study of its method of government. If at any stage the social organisation and economic control faltered, every element in such a complex system would have broken down.

Great Zimbabwe is the principal of the country's 140 historical monuments and sites, all administered by National Museums and Monuments of Zimbabwe. Among them are Stone Age and Iron Age sites, places of scenic beauty and rock paintings. Zimbabwe is very rich in rock art, with an official listing of more than 3,500 sites with prehistoric paintings on rock faces, overhangs and shelters.

Opposite and above:
Great Zimbabwe, once the largest city in sub-Saharan Africa, became wealthy as a trading centre between the goldfields of the interior and the east coast.

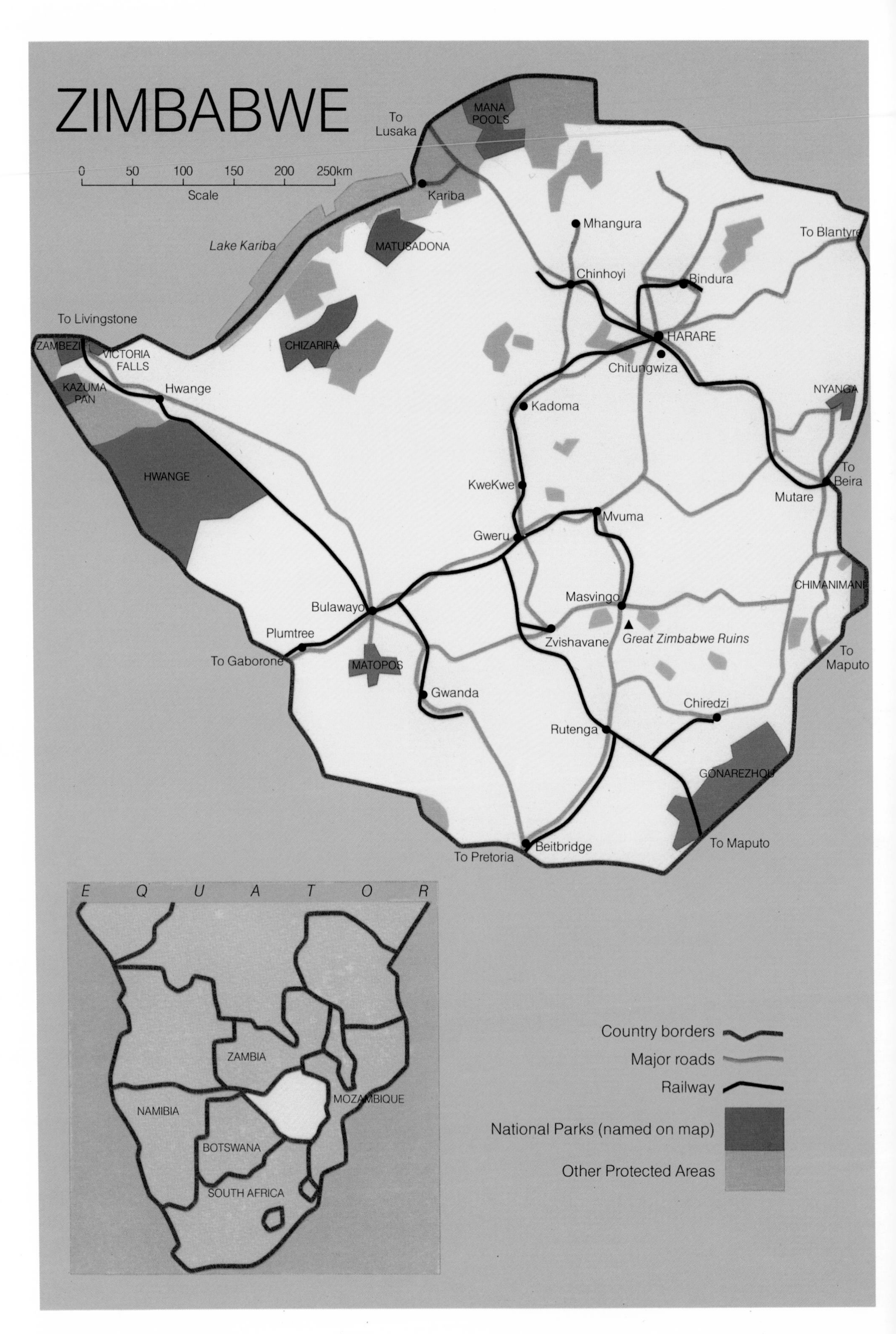
ZIMBABWE
0 50 100 150 200 250km
Scale
To Lusaka
MANA POOLS
Kariba
Lake Kariba
MATUSADONA
Mhangura
To Blantyre
Chinhoyi
Bindura
To Livingstone
ZAMBEZI
VICTORIA FALLS
CHIZARIRA
HARARE
Chitungwiza
KAZUMA PAN
Hwange
Kadoma
NYANGA
HWANGE
KweKwe
To Beira
Mutare
Mvuma
Gweru
CHIMANIMANI
Masvingo
Bulawayo
Plumtree
Zvishavane
Great Zimbabwe Ruins
To Gaborone
MATOPOS
To Maputo
Gwanda
Chiredzi
Rutenga
GONAREZHOU
Beitbridge
To Pretoria
To Maputo
E Q U A T O R
ZAMBIA
NAMIBIA
MOZAMBIQUE
BOTSWANA
SOUTH AFRICA
Country borders
Major roads
Railway
National Parks (named on map)
Other Protected Areas

Geography

Landlocked Zimbabwe is the southernmost country in tropical Africa, with a north-south axis of 720 kilometres and an east-west axis of 830 kilometres. It has an outstanding natural heritage.

The contrast is extraordinary – from rich farmland to desert scrub and from modern city sprawl to rustic village cameos, little changed in more than a century.

Temperature ranges add to the variety. There are summer highs of nearly 40 degrees centigrade in the low-lying areas and, in the winter, sub-zero highveld frosts. Rainfall also varies enormously, from over 3,000 mm a year in the eastern mountains to 300 mm in the most arid lowveld. Consequently, vegetation ranges from thick evergreen forest to drought-resistant thorn scrub.

Zimbabwe sits on the map like a compact teapot, bordered on all sides by Zambia, Mozambique, South Africa and Botswana; and with the westernmost tip of the 'spout' (where four countries meet) butting onto the Caprivi Strip of Namibia. The country is bordered along much of the north and south by the Zambezi and Limpopo rivers.

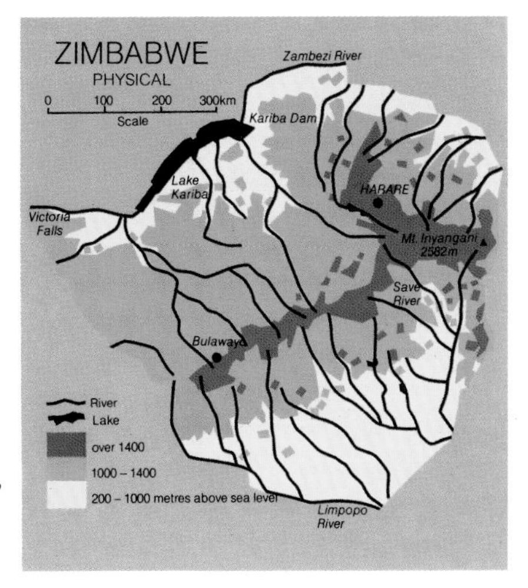

There are four distinctive land types – what the geographers call physiographic regions. These are the Zambezi Valley in the north; the dominant plateau wedge of the Midlands, fanning out from south-west to north-east; the Save-Limpopo lowlands in the south; and the Eastern Highlands. Within these areas, the altitude ranges from less than 250 metres above sea level to over 2,600 metres.

The Eastern Highlands, bordering Mozambique, is a belt of mountains and high plateaux which marks the edge of the great tableland of south-central Africa. From north to south, running for about 250 kilometres, are the Nyanga Mountains, the Vumba Mountains, the Melsetter Upland and the Chimanimani Mountains. It is a beautiful and productive ridge of high country, with refreshing scenery and crisp mountain air.

Moving south-westward, from the craggy Chimanimani Mountains, the countryside falls away to the Save River, which carries the broadening scars of siltation caused by severe soil erosion in the area. Here, the dry, flat landscape of the south-eastern lowveld is dramatically broken by large green mosaics of irrigated sugar cane, wheat and cotton.

In the extreme south-west, near Bulawayo, is the narrowest and lowest part of the great plateau that dominates the central part of the country. Here is the huge granite tumble of the Matopo Hills, a mass of rocky outcrops sculptured over millennia into great domes and crenellated ridges surmounted by remarkable 'balancing rocks'.

For the most part, western Zimbabwe is featureless and flat; sadly, axes and chainsaws have taken their toll of once-great natural 'dry' forests of Zambezi teak, mukwa and panga-panga hardwoods. The railway line to Zambia takes advantage of this monotonous terrain and runs arrow-straight for 112 kilometres between Gwayi and Dete. Most of the track runs along the eastern boundary of Hwange National Park – one of the best known game reserves in Africa. Also nearby are the coalfields of Hwange, where huge reserves of the 'black stones that burn' (as the local people described coal to early prospectors, at the turn of the century) are mined deep underground and with extensive surface stripping.

North-west of Hwange is Victoria Falls – known locally as Mosi oa Tunya ('The Smoke that Thunders') – named by missionary-explorer David Livingstone, in honour of his queen, when he became the first European to see them in 1855. The largest single curtain of falling water on earth, Victoria Falls on the Zambezi river draws visitors from all over the world. The great waterfall is a unique geological feature; each of the seven dramatic downstream gorges was once the lip of the falls, the erosive power of the water continually cutting back into the traverse faults that criss-cross the basalt rock of the river bed. At Devil's Cataract, the western edge of the falls, another fault is being excavated by the surging water – and will eventually form the new lip.

Downstream, the winding ribbon of the Zambezi feeds Lake Kariba, formed by the damming of the river between 1955 and 1959. One of

Harare is an attractive spacious city with modern buildings and wide, tree-lined streets.

the largest man-made lakes in the world, Kariba generates hydroelectric power for Zimbabwe and Zambia, is a major tourist attraction and supports a growing commercial fishing industry.

To the south-east of Kariba is the Great Dyke – the most outstanding feature on the geological map of Zimbabwe. Running roughly north-south for about 530 kilometres, and varying in width from one to eleven kilometres, this spine of hills bisects the central plateau. It is the longest linear rock mass in the world. Although its serpentine soils are toxic to most plants, they nourish more than 20 species which grow nowhere else in the world. The Dyke is also of great economic significance, since it holds enormous deposits of chrome and platinum.

Not far from the Great Dyke is Harare, Zimbabwe's modern capital city which, with the satellite town of Chitungwiza, has a booming population of nearly a million people. The natural woodland which, even ten years ago, blanketed the urban fringe has been depleted. But, despite the city's continuing rapid growth, there is very little of the shanty sprawl that characterises so many cities in the developing world.

Zimbabwe at a Glance

Total area:	390,580 sq km (150,804 sq miles);
Neighbouring countries:	Botswana, Namibia, Zambia, Mozambique, South Africa;
Population:	8,870,000 (mid-1988 estimate) with just over three per cent annual growth rate;
Urban population:	1,770,000 (mid-1988 estimate)i.e. 20 per cent of total;
Major towns:	Harare (capital), Bulawayo, Chitungwiza, Gweru, Mutare, Kwekwe, Kadoma, Masvingo;
Languages:	English (official); Sindebele and Chishona predominant;
Religion:	Tribal beliefs, Christianity;
Government:	The Republic of Zimbabwe has a participatory democracy form of government, headed by an Executive President, Robert Mugabe, who has been in office since 1980 as Prime Minister until 1988 when he became President;
Labour force:	34 per cent in services; 27 per cent in agriculture; 16 per cent in manufacturing; five per cent in mining; 18 per cent other (1985).

A dramatic range of landscapes greets the visitor to Zimbabwe. Although in the main these are a product of specific rock types, climate and altitude conditions, the influence of man is becoming increasingly important. From top to bottom: Matopos Hills, Hwange mopane woodland, the Great Dyke and view over the Honde Valley in the Eastern Highlands. (DR, DH, DR, SB)

DHLIWAYO'S BUTCHERY
UNSURPASSED MEAT.
MATAPI HAIRDRESSING
SALOON
NHEKAIRO MACHIPISA
STORE & RESTAURANT
When
you know
MUSAMI-DARANGWA

People of the Land

Visit a bus station in any urban centre, on any weekend of the year, and you witness a colourful and cardinal feature of Zimbabwean society – the deep rural roots of the people.

Buses bound for just about every corner of the country will have their full-length roof-racks laden with goods from the shops and factories of the city, for the rural homes and families of their passengers. On the return journeys, the same roof-racks are laden with produce from the rural areas, for sale in the city marketplace.

Zimbabweans are essentially a rural people, with nearly three-quarters of the population living either in the communal lands (54.3 per cent of the country's area) or on commercial farming and state land (20 per cent). Only one in four Zimbabweans live in towns and cities – and very few of these forget their rural roots. The urban businessman, the academic, the professional, the civil servant and the politician all see their true homes as somewhere in the country, well away from their suburban residences and city lifestyles. The concept of a landless urban proletariat has only just begun to evolve.

Mobility between town and country is very high. The wide network of bus routes (which are operated not as a subsidised social service but by private enterprise) enables people to travel on a regular basis, even to and from the remotest areas.

The people of town and country are closely interdependent – a third of the households on communal land depend, at least in part, on income from family members working in the towns and cities. Urban areas are the prime markets for agricultural produce – and the main source of goods and services. In turn, the rural areas provide food, temporary labour and social security for townspeople, who return to their homes in the country at the end of their working lives to be fed and sheltered in their old age. This security system, based on the strong kinship network, is very important to the great majority of workers, who are not covered by any national welfare system.

This closeness to the land is at the heart of indigenous cultures in Zimbabwe. Historically, the people's perspective of their lives and their nation is wrapped in rural idiom. This finds expression in oral tradition, in song and dance, and in a variety of customs. It was also a powerful bonding factor during the long struggle for political independence, when one of the phrases of the common people was *vanhu ve'ivu* – which, in Shona, means 'people of the land'.

It is therefore perhaps ironic that the biggest single factor that has shaped the face of the Zimbabwe of today was the division of land on racial and other lines, that took place progressively after the British occupation of the country in 1890.

Gold was the strong lure that first brought white settlers to Zimbabwe – but the chance to own a large piece of farmland was also a powerful attraction. After the occupation of Mashonaland in 1890, the British South Africa Company (BSAC) awarded land grants of 1,210 hectares, and 15 gold claims, to each man in the first group of settlers. A further grant of 2,420 hectares each was given to those who took part in the war with the Ndebele a few years later.

This casual parcelling of land became formalised with the establishment of the Gwaai and Shangani Reserves, in the west, for the defeated Ndebele. But with the quelling of the Ndebele and Shona 'uprisings' in 1896-97, it became apparent that the indigenous people were in danger of losing large parts of their land. The British government then insisted that the BSAC set aside areas specifically for them. So began the designation of 'native reserves' throughout the country. Later termed 'Tribal Trust Lands' and now known as communal lands, these areas make up 42 per cent of Zimbabwe.

This division of land set the pattern for all subsequent land use, both good and bad. It was the situation inherited by the new administration, at independence, some 90 years later. Today, a large measure of environmental degradation – overgrazing, deforestation and soil erosion, in particular – is due to this historical concentration of the African population on poor land.

Opposite:
Zimbabwe's urban dwellers maintain close links with their rural families.

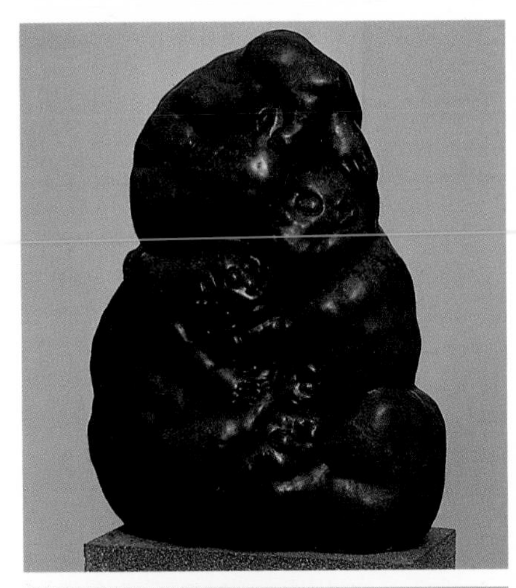

Above right:
'Rhino Totem: Father and Child' by Bernard Matemera. International art critics consider several Zimbabwean sculptors to be amongst the best in the world today. (Collection of the National Gallery of Zimbabwe).

Above:
Two sculptures by Thomas Mukarobgwa: 'Family Dreaming' (top) and 'Father, where is Mother?' (bottom). (Collection of the National Gallery of Zimbabwe).

Inspiration for a Sculptor

Sculptor Bernard Matemera is one of several Zimbabwean stone carvers whose work is winning international acclaim for its originality and strength. He obtains his inspiration working under a wild fig tree, outside his home at the Tengenenge sculpture community in Guruve. As the day wears on, he moves his stool between two or three heavy pieces of stone, which take shape together, to make sure that he is always working in direct sunlight.

Finished and polished, these dream-inspired human-animal forms, grotesquely powerful in their gleaming roundness, find their way into leading galleries, exhibitions, collections and homes in many parts of the world.

Bernard Matemera has worked the lustrous black, grey and deep red serpentine for more than 20 years. He is a quiet man, with the grip of a blacksmith, and has been a major force in establishing Zimbabwean sculpture in the art world. He is among a handful of Zimbabwean carvers whose work is promoted (at home and abroad) by the National Gallery of Zimbabwe, which has been a prime catalyst for all forms of indigenous art over the years.

The gallery also displays and markets the best of other

Traditional Zimbabwean craftwork.
From left to right:
Ceramic pots, wooden stools, chairs and headrests, Gudza dolls, drums made from wood and animal skins, baskets for a variety of uses, and a Makishi dance costume made from bark fibre cloth on a wicker frame.

traditional Zimbabwean craftwork which, with its different cultural origins, varies greatly in style and technique. Made from the rich and diverse natural products of the earth – such as clay, stone, wood, grasses, fibres and gourds – they include pots, baskets, mats, wooden implements, toys and musical instruments. Art and craft centres in different parts of the country provide materials and tuition to develop local talent and to promote sales.

Traditional woodcarving includes walking-sticks, stools, headrests, ceremonial axes and many ritual objects, such as the divining 'bones' of the traditional healer. There are also some unique fibre blankets, hats and skirts. These are woven, without a loom, from strands of bark (called gudza) from the musasa tree *(Brachystegia spiciformis)* and similar species.

Many of the items are artistic as well as functional. Pottery for cooking, brewing and storing has colour designs, using black graphite and other natural pigments. The baskets – which are often made of ilala palm *(Hyphaena spp)* and come in many shapes and sizes – carry traditional patterns worked with strands dyed dark brown with bark, or grey and black with mud.

As with so many Zimbabwean crafts, their patterns and shapes evoke the things of everyday life and nature – such as the Milky Way, hills, trees and animals.

Economics

Not yet a decade old, Zimbabwe is Africa's newest independent nation. Yet it plays a leading role on the African continent. It is a prominent and productive member of the Southern African Development Coordination Conference (SADCC) and the Preferential Trade Area (PTA) groupings of African countries. It also has the resources and the capability to contribute significantly to trade in these regions – and beyond – with well-developed transport and communications systems.

Zimbabwe's economy is one of the most extensively diversified in the developing world. It is based on an abundant supply of mineral and agricultural raw materials, a highly developed infrastructure, a relatively skilled labour force, and experienced management.

The Government, local authorities and public corporations account for about 30 per cent of the Gross Domestic Product. The manufacturing sector contributes about 25 per cent; agriculture fifteen per cent; distribution fifteen per cent; and mining six per cent. Although the primary sectors (agriculture and mining) together account for only about 21 per cent of GDP, they provide over 70 per cent of the country's total merchandise exports; the remainder comes from the manufacturing sector.

Exports are dominated by primary products derived from natural resources: tobacco, gold, ferrochrome, cotton, asbestos, nickel, steel, sugar, copper and meat. The major imports are liquid fuels (20 per cent of the total) and capital and intermediate goods.

As a leading member of the 101-nation Non-aligned Movement (President Mugabe was its 1988 Chairman) Zimbabwe is intent on restructuring its trade links. The emphasis will be on moving away from excessive dependence on the OECD countries of Western Europe, North America and Japan, and away from South Africa, towards other countries in Africa, the Middle East and the socialist economies of Eastern Europe, Asia and Latin America.

The Government is committed to a policy of 'Growth with Equity' which is designed to redistribute income and wealth (including land) and to ensure equal opportunities for everyone, with particular respect to education, health services and employment. There is a substantial, and growing, degree of government participation in the economy. It is also official policy to reduce the extent of foreign ownership and control over the economy.

But resilient, productive and broad-based though it is, the Zimbabwean economy is under severe strain. There has been limited growth in recent years, with rapid inflation, falling investment and escalating unemployment. The balance of payments deficit is high. Heavy external debt-service obligations, the sluggish world economy, three severe droughts since 1980, high government spending at home, serious foreign exchange shortages and depressed demand and prices for primary commodity exports (on which Zimbabwe depends so heavily) have combined to put tight constraints on economic growth.

Unemployment is perhaps the most serious problem facing Zimbabwe in the 1990s. Confronted with a sharply rising tide of school-leavers and graduates, as a result of the dramatic expansion of the education system since independence, the planners seem caught between the demands of social transformation and egalitarianism on the one hand, and the imperatives of rapid economic growth on the other.

Zimbabwe's five-year National Development Plan (to 1990) shows the Government's commitment to natural resource conservation as an integral part of economic reform. The Plan has six key objectives:

- the transformation, control and expansion of the economy (through investment by the state, cooperatives and the private sector, separately or in joint ventures) so that a greater proportion of it is owned and controlled by Zimbabweans;
- land reform and more efficient utilisation of the land; the Government is giving priority to land reform by increasing the number of state farms, intensifying resettlement schemes and promoting cooperatives;
- raising the living standards of the people; special efforts are being made in rural areas by increasing agricultural productivity;

Sugar cane is burned to make it easier to harvest. Sugar ethanol also burns well, in a 15:85 admixture with petrol in car engines. Started in 1982, Zimbabwe's ethanol plant produces 40 million litres a year, saving the foreign exchange costs of an equivalent volume of petrol.

- increasing employment and the development of manpower; the Government is expanding the infrastructure and institutions necessary for a co-ordinated national training programme;
- the development of science and technology;
- the maintenance of a correct balance between development and the environment.

The urgent challenge facing Zimbabwe is to find – and implement – ways of boosting manufacturing exports (which are not subject to the vagaries of the international commodity markets) and of attracting more investment to generate economic growth.

The challenge is formidable. But Zimbabwe is better placed than most sub-Saharan countries to meet it – if only because its renewable natural resource capital has enormous productive potential.

Zimbabwe's exports are dominated by primary products derived from natural resources such as asbestos.

Chapter Two

STATE OF THE WILD

A Wealth of Wildlife

Unlike many countries in Africa, Zimbabwe still has a rich variety of indigenous wildlife: including some 270 species of mammals, about 640 different birds and 153 reptiles.

Where suitable habitats remain, many species have survived in reasonable numbers. The bat-eared fox, an inhabitant of the more arid western parts of the country, has actually extended its range in the Hwange-Victoria Falls region. Some species, such as the African wild cat, duiker and many birds, have adapted to living on farmland or close to human settlements. Even hippos sometimes turn up on the outskirts of urban areas. A number of other species, notably the chacma baboon, bushpig, red-billed quelea and several others, have been so successful that they are now considered to be agricultural pests.

Elsewhere the cheetah, although probably never a numerous animal in Zimbabwe, seems to be increasing in numbers, particularly in the southern half of the country. There has also been a growing number of sightings in the Zambezi Valley and Matusadona National Park, though this may merely reflect an increasing level of tourism in these areas.

Several mammals do, however, give cause for concern. Although well represented in neighbouring countries, Lichtenstein's hartebeest is now in serious danger of extinction in Zimbabwe, with only 50 to 70 survivors, living in a small area in the south-east. There are, however, plans to develop further populations of this antelope, initially within Kyle Recreational Park near Masvingo.

The survival of the wild dog, or Cape hunting dog, is even more in the balance. After the black rhino, this beautiful social carnivore is Zimbabwe's most endangered mammal. For many years, it was seen as a threat to both cattle and other wildlife – and merciless persecution drove it to near extinction in the space of a few decades. Common sense finally prevailed, but not until the national population had fallen to about 400 animals. A captive breeding programme forms part of recent plans to introduce stringent conservation measures for the species.

There are many interesting birds in Zimbabwe. Two species, in particular – Swynnerton's robin and Robert's prinia – are found nowhere else in the world. They are confined to forests in the eastern districts. Neither is threatened, but experts are concerned that habitat destruction in the region may lead to future problems. There are also some very well-represented bird families here, notably the warblers, with 53 species, and diurnal birds of prey, with an outstanding 47 species.

No birds have actually become extinct in Zimbabwe in recent years – but many, such as the bateleur eagle and a number of other raptors, have declined in numbers or disappeared from some of their original haunts. Human activities, such as deforestation and continued use of persistent pesticides, are largely to blame.

Relatively little is known of the status of Zimbabwean tortoises, terrapins, lizards, snakes and other reptiles.

The best-known species is probably the Nile crocodile, both

Opposite:
A herd of buffalo in the Zambezi Valley, an area of exceptional species diversity. (PB)

Top and above:
Zimbabwe has a rich variety of wildlife, including this beautiful white-faced owl, photographed near Harare, and the minute hatching bugs. (DH)

because of past persecution and because of its present economic value. At one time, it had been virtually exterminated across much of Zimbabwe. But the development of crocodile farms to supply hides to luxury fashion markets is helping to ensure the survival of the species. Crocodile farmers were formerly obliged by law to return a percentage of their hatchlings to suitable areas in the wild, where populations had decreased or been eliminated. Today, however, the Zimbabwean shore of Lake Kariba is thought to have a population of over 30,000 crocs, in some areas at a density of one mature animal for every 200 metres of shoreline. Major rivers such as the Zambezi, Save, Runde and Mwenezi also have good crocodile populations.

The more conspicuous snakes, such as the Egyptian cobra, black mamba, puff adder and boomslang, are inevitably a focal point of interest. They all pose a potential threat to unwary people, and often occur near human habitation. Zimbabwe's most spectacular snake, the python, is currently included on the Specially Protected Species list, because of past demand for its skin. Nevertheless, it is fairly numerous today, both inside and outside protected areas.

Animals on the Specially Protected Species list are excluded from the general provision of the 1975 Parks and Wildlife Act giving landowners jurisdiction over wildlife on their land. The python is the only reptile on the list – but a number of mammals and birds have been included. They are not necessarily all threatened species; some are there for cultural reasons, others as a precautionary measure until more is known of their status and distribution.

Wildlife legislation here is both unusual and enlightened, with two dominant features. Firstly, the 1975 Parks and Wildlife Act gives landowners control over the use of wildlife on their land, through such activities as ranching, tourism and safari hunting; this provides an incentive to manage wildlife populations carefully. Secondly, all visitors to protected areas administered by the Department of National Parks and Wildlife Management are carefully controlled.

These two factors, coupled with the high level of awareness, within central Government, of the value of the Department (and a consequent full measure of support for its activities) augur well for the continued effectiveness of wildlife legislation in the years ahead.

Rhinos and Elephants

With such tremendous public appeal,these two larger-than-life animals have assumed 'flagship' roles for wildlife conservation in Africa. Both rhinos and elephants carry valuable products and are therefore highly valued themselves. But in Zimbabwe, they present sharply contrasting management problems – too many elephants which have to be reduced (very profitably) from time to time; and among the last significant black rhino population on the continent, requiring intense protection against the onslaught of well-armed poachers.

At the turn of the century, black rhinos were widely distributed in Zimbabwe. But with expanding human populations and agricultural development their range was gradually reduced until, by the 1920s, they had disappeared altogether from the western parts of the country; over the next 30 years, they disappeared from other areas as well.

By 1960, their refuge was the Zambezi Valley, though a few animals still survived in the Chipinge area close to the Mozambique border. Following the filling of Lake Kariba in the same year, some of the Zambezi animals were reintroduced to Hwange National Park and, in 1972, the Chipinge population was moved south to Gonarezhou National Park. Capture operations during the mid-1960s, and from 1972 to 1976, also resulted in the relocation of animals, from communal lands into Hwange, Matetsi, Chirisa and Gonarezhou.

During the 1960s, rhinos were poached for meat. At the time, there was little evidence of poaching for horn. A decade later, however, the price of horn rose rapidly – largely due to oil price rises and thus the growing wealth of men in North Yemen, where the horns are fashioned into dagger handles. Nevertheless, military activities during the civil war

dissuaded attempts at poaching in Zimbabwe. Besides, there were larger numbers of more accessible rhinos in other African countries.

But by 1983, the once-thriving black rhino population in the Luangwa Valley in Zambia had been greatly depleted. All attention was suddenly focused on Zimbabwe – there was no longer a civil war and its black rhino population was still a healthy 1,700.

The inevitable onslaught from across the Zambian border began in December 1984. In response, the Department of National Parks and Wildlife Management quickly launched a survival programme for the animals. Many international organisations, in particular USAID, SAVE and WWF, have since rallied in support; and the Zimbabwe Rhino Survival Campaign was specially set up, under the auspices of the Zimbabwe National Conservation Trust. The help from all these sources has included a sophisticated (and highly effective) radio network for the valley, the provision of vehicles and staff housing, a helicopter patrol and many items of equipment.

'Operation Stronghold' is the name given to these efforts to protect black rhinos in the Zambezi Valley. The name is appropriate. Zimbabwe really is one of the last strongholds for the black rhino in the wild in Africa. The species has declined in numbers from 65,000 in 1970 to less than 4,500 today. Zimbabwe holds about half of all these animals – and the Zambezi Valley has the only contiguous population of more than 500 on the continent.

There, in the heat and thick bush, a concerted last-ditch stand is being made against the operations of the Mafia-like poaching organisations. Armed with hunting rifles to kill the rhinos and automatic assault rifles to resist capture, the poachers are determined. Their job is to feed an insatiable and misguided demand for rhino horn, notably from North Yemen, where it is used to make dagger handles and from east Asia, where it retains a reputation as a medication.

But most of the profits – which are enormous – go to middle-men and the main organisers of the poaching operations. For comparatively paltry rewards the poacher on the ground risks his life – and often loses it. More than 30 poachers were killed (and nearly as many captured) by Zimbabwean anti-poaching units between the beginning of 1985 and the

The Department of National Parks and Wildlife Management is mounting a very professional campaign against rhino poaching. Nevertheless, hundreds of rhinos are still being slaughtered – leaving young orphans which do not yet possess the prized horn. The Department carefully selects farms to act as rhino 'orphanages'. Here, Norman Travers and Godfrey Charakupa of Imire Game Park feed baby black rhinos which have been fostered. White rhinos were once widely distributed in Zimbabwe, but were hunted out by 1934. Following their reintroduction in 1960, they are now restricted to Hwange, Matopos, Lake Kyle and Lake McIlwaine National Parks and a few private farms.

Zimbabwe's thriving elephant population – of around 40,000 – has been built up by good management. In the early days of British colonisation, excessive hunting had reduced the population to fewer than 4,000.

end of 1987. In the same three-year period, the poachers slaughtered about 380 rhinos.

As the day-to-day confrontations continue in the Valley, longer-term strategies to safeguard the animals are being developed. These include establishing breeding units in other protected areas – more than 120 Zambezi animals have so far been relocated in Hwange National Park, for example – and on some of the larger ranches in the country; some animals will also be kept in intensive captive breeding centres, both in Zimbabwe and in North America.

Some argue that captive breeding programmes and anti-poaching patrols, while necessary, are not getting to the core of the problem. The ultimate solution really lies in controlling the worldwide corruption that makes illegal international trade in rhino horn possible.

It is precisely the same approach that will be required to control elephant poaching. Unlike in rhino products, a certain amount of trade in ivory is allowed on the world marketplace. But of 100,000 tusks which appeared in the international trade in 1986, no less than 75,000 lacked legal documentation.

Africa's elephant population currently stands at about 750,000. Largely due to poaching, it is declining at the rate of about ten per cent every year. Elephant poaching is less serious in Zimbabwe than in other parts of Africa but, even here, it is sometimes intensive in the south-western corner of Hwange National Park and in Gonarezhou National Park. The poachers are mainly from neighbouring countries.

In some countries, elephants are seriously threatened but the irony is that, in others, they are almost over-abundant. In Zimbabwe, for example, elephants are a real asset – and a prime example of the sustainable utilisation of a natural resource. Their value is fully recognised and this abundance is largely thanks to careful management of the elephants as an exploitable natural resource.

But they must 'earn their keep' in return for this enlightened form of conservation. They bring in considerable foreign exchange from safari hunting and tourism; their ivory and hides fetch good prices at government auctions, following control operations; they provide large quantities of meat; and they put development money back into the local communities.

About 100 bulls are placed on a quota each year for safari hunters and communal area hunting concessions. Each animal is worth about Z$10,000 in ivory, hide and meat. On top of this, there is a trophy fee of Z$5,000 and, since an elephant can be hunted only on a fifteen-day or 21-day safari, a significant income from other aspects of the trip.

The records of Portuguese trading posts on the Mozambique coast show that ivory has been harvested from this part of Africa for at least five centuries. The precise effect on the elephant population is not clear. In Zimbabwe, there were probably fewer than 4,000 left at the turn of the century. But very strict controls on hunting allowed the population to recover and, by 1960, it had increased once again to more than 30,000.

There was, however, an adverse result of this increase, particularly in the national parks. The elephants were doing serious damage to their woodland homes. By the mid-1960s the authorities were sufficiently concerned to recommend that some elephant populations be reduced. Recent studies have now shown that densities of over one elephant per square kilometre are likely to lead to accelerated woodland damage. In some cases, where elephant densities have been high, a tree mortality of more than 20 per cent a year has been recorded.

Outside the parks, there has been a continuing expansion of human settlement – with increasing conflict between villagers and elephants. This has both reduced the range available to the animals and led to the killing of large numbers for crop protection and for tsetse fly control.

For all these reasons – from sports hunting to disease control – in the 20-year period from 1960 to 1980, about 18,000 elephants were killed. Yet their numbers continued to rise and topped 50,000 by 1982.

Over the past few years attempts have been made to reduce overall elephant densities to less than 0.8 animals per square kilometre within protected areas. This would result in an elephant population of about 34,000 within these areas. The total is still more than 40,000 – and the annual growth rate in Zimbabwe is currently about five per cent.

Zimbabwe's healthy elephant state is the result of good management over many years. An important component of this management is a keen knowledge of the animals themselves. There have been many detailed studies over the years – for example, on their patterns of movement both within and outside protected areas. In particular, for the past two decades, elephant populations have been regularly counted using aerial surveys conducted by the Department of National Parks and Wildlife Management. Such population data is also used in computer-based models, which are an important tool in balancing elephant numbers with the availability of natural resources.

These data are crucial in successful elephant population management, particularly in the setting of quotas for the sustained yield of high quality trophy bulls, which is so important to the safari industry. But it is nevertheless a fine balance which has been so carefully achieved – and must be maintained in the long term – between elephant abundance and destruction of the woodland that supports them.

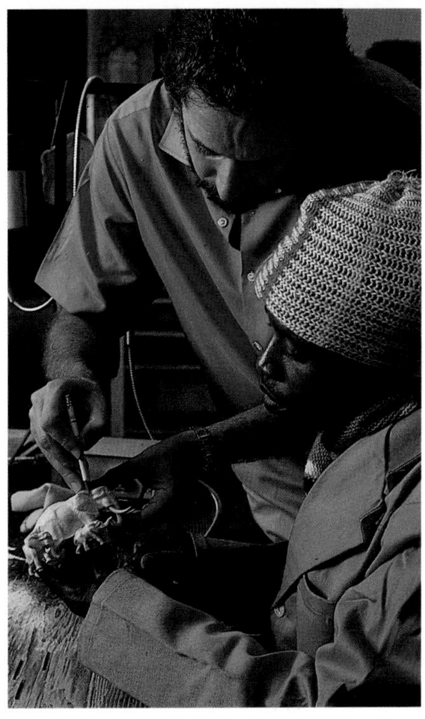

Top:
An essential part of elephant management is population control. With elephants confined to limited areas, a general maximum of one animal per 1.25 square kilometres must be ensured. Otherwise, elephants begin to destroy their own habitat. A useful byproduct of culling is the meat, hide and ivory – one elephant could be worth up to US$5000 in raw materials alone. (HC)

Above:
The Zimbabwean Government ensures that the country receives the maximum benefit from legitimate ivory, by selling raw ivory through a well-organised auction system. Highly skilled artists, such as Patrick Mavros and his assistants, add great value to the raw product by fashioning exquisite carvings.

Top:
From scratch, Tom Muller worked with nature to create many kinds of African woodlands, forests and other habitats – all in a 68 hectare plot in central Harare.

Middle and above:
Zimbabwe has more than 100 plant species, including Aloe ortholopha *(middle), which are found nowhere else in the world. The sabi star (above) is among the country's rich indigenous flora.* (MK, DH)

An Investment in Plants

There are an estimated 5200 different species of plants in Zimbabwe. More than 100 of these are found nowhere else in the world. Yet relatively little is known about this rich natural heritage and, as botanist Tom Muller points out: 'You can't conserve without knowing what you have.'

Twenty-five years ago Swiss-born Tom Muller started creating, from scratch, the 68-hectare National Botanic Garden in central Harare. He was one of the first people in the world to arrange plant species in ecological groups rather than in more traditional taxonomic or geometrical formats. Beautifully sculptured around the original *Brachystegia* woodland, amid rolling lawns which are progressively diminishing as the plantings mature, the Garden is now regarded as among the finest in Africa.

Tom Muller, Curator of the National Botanic Garden and Herbarium, regards the task as half complete. But he is also concerned with other aspects of plant research and conservation in Zimbabwe. In particular, he welcomes the current communal lands natural resources survey, which follows a detailed vegetation survey recently completed in the national parks. He hopes this will soon lead to a comprehensive study in the remaining parts of the country.

In Zimbabwe, large areas are still covered by natural vegetation, such as grassland and woodland. So a complete and accurate picture of its composition would be a significant help in land-use planning and would therefore go a long way towards making better use of the vegetation.

Locating the presence of any threatened (or otherwise noteworthy) species is an important aspect of any survey work. We cannot risk losing a single species – because we rely on plants in every field of life. They protect the soil and prevent land from becoming desert; they help to maintain freshwater supplies; they are essential features of habitats; they provide the medical products which cure even the most serious illnesses; they feed us and clothe us; they provide us with fuelwood and even with building and industrial materials.

The loss of any one of Zimbabwe's endemics, for example, would mean the loss of irreplaceable genetic material which has taken millions of years to evolve. There are two known areas of endemism in the country. One of these is the Great Dyke, with more than 20 plant species which are found nowhere else in the world; among them are three succulents *(Euphorbia memoralis, Euphorbia wildii* and *Aloe ortholopha)* that are much sought-after by collectors worldwide. The other area is in the rugged and remote Chimanimani Mountains, with more than 40 endemics thriving on the poor quartzite soils there.

As far as we know, none of them is particularly threatened. But the need for some form of protection, particularly in the more accessible Great Dyke area, is evident. A potentially outstanding site for a nature reserve at the Dyke has already been spoilt by co-operative chrome mining activities.

Zimbabwe is not unusual in lacking the knowledge which is necessary for a definitive plant conservation programme. World-wide, scientists often do not know which species are endangered – and, even if they do, the lack of information on them is often so acute that their distribution is virtually unknown.

But already it is clear that there are small areas of vegetation, throughout Zimbabwe, which need protection and which the country could afford to set aside. As Tom Muller says: 'It is not asking much, and it would be a sensible investment for posterity.'

Parks and Reserves

Few developing countries set aside more than twelve per cent of their area purely for wildlife protection and human enjoyment and education. But Zimbabwe is an exception. Its Parks and Wildlife Estate totals 46,000 square kilometres and includes every ecological region of the country. The reason is simple – the area is ecologically very rich, and tourism revenues are high.

The Estate takes in dramatic landscapes, great sweeps of bush, mountain and forest, rivers, lakes and waterfalls; it also includes botanic gardens and other relatively small areas of special interest. It manages some of the largest and most varied concentrations of wildlife in Africa, and focuses right down to the protection of individual plant species that grow nowhere else on earth. It embraces the world-famous as well as the virtually unknown. And it is the basis of a major tourist industry. Administration of

Above left:
An intimate knowledge of wild species – and their relationships with one another and with their habitat – is essential for wildlife conservation and utilisation. The Zimbabwean Government therefore attaches a high priority to ecological research. (DP)

Above:
The klipspringer lives in rocky habitats and rarely strays from its home on hillsides, ridges or kopjes. Like other dwarf antelopes, the scent glands in front of the eyes are very well developed. (DP)

the Estate is an enormous task: the Department of National Parks and Wildlife Management is the largest single land-use agency in the country.

There are three distinct areas of activity within this fragmented and varied natural realm. Special Conservation Areas (to which access is strictly controlled) are generally small and designed to protect particularly sensitive biological features or to conduct research. Wilderness Areas are open to a limited number of people and no development is allowed. Finally, Wild Areas, which are open to more people, allow the building of roads, game-viewing facilities and visitor accommodation. It is estimated that the level of development could be increased – to cater for five times the present number of visitors – without serious impact on the Estate's natural qualities, or a lowering of the present high aesthetic standards. On the ground, the Estate consists of six different types of protected area:

- National Parks: Hwange, Victoria Falls and Mana Pools are the best known; there are eleven altogether;
- Safari Areas: a total of sixteen; they provide a wide range of activities besides hunting – including hiking, angling, game viewing and photography;
- Recreational Parks: these are generally associated with major dams, such as Kariba, Kyle and McIlwaine, and receive large numbers of visitors;
- Sanctuaries, Botanical Reserves and Botanical Gardens: these protect individual animals, plants or biotic communities and, in the case of botanic gardens, propagate mainly Zimbabwean species or particular groups of plants such as aloes and cycads.

The Parks and Wildlife Estate thus serves many purposes – commercial, environmental, research, aesthetic and educational. It is an exciting open-air classroom for the young to learn about conservation and the wise management of natural resources. It also provides an example of good conservation practices for other sectors in Zimbabwe to follow. This is vitally important to the future of Zimbabwe's protected areas. In the years to come, they will undoubtedly come under frequent scrutiny by proponents of conflicting forms of land use. They will need the support of people who understand the social and economic importance of conservation in general, and of Zimbabwe's protected areas in particular.

Wildlife Leadership

When he was a boy, Willie Kusezweni Nduku and his playmates used to hunt with bows and arrows. They regularly stalked the bush and forest around their village in Mount Selinda, near the Mozambique border in south-east Zimbabwe. After a day of sport, drinking and bathing in mountain streams, they would return home with their trophies for the family pot: a hare, a duiker, perhaps an impala. Once, young Willie brought down an adult kudu.

Now, as Director of National Parks and Wildlife Management, Dr. Nduku is directly responsible for an area totalling more than twelve per cent of the country, including the Zimbabwean portion of Lake Kariba – and he has a staff of some 4,000. It is a responsibility which he wears as comfortably as the open-necked khaki shirt that makes up part of the National Parks uniform.

'We in Zimbabwe have much to be proud of – and grateful for – in our parks and wildlife areas,' he says. 'They are recognised as important models in other parts of Africa. But we are now facing the biggest challenge of all. We have to spread the conservation ethic among the mass of the people in the communal lands. They must be encouraged to take responsibility for wildlife management in their areas; they need to be made accountable for the resource – in return for directly benefiting from it. That will demand major changes in traditional attitudes and practices and there is much work to be done to bring this about.'

Dr. Nduku believes that, with patience and application, there is a good chance of success – provided the Forestry Commission can establish sufficient woodlots throughout the country to halt the accelerating destruction of natural woodlands. If the trees go, he observes, everything else will go too.

He is particularly concerned about the future of the middle Zambezi Valley where, until now, the only form of protection for wildlife and the environment has been the tsetse fly.

'With the progressive elimination of the fly, human settlement must take place only after careful study and planning. If people just move in with their cattle the land will be quickly degraded and we stand to lose a large and very special part of the country. Good coordination and strong control are vital.'

The boy who drew his bow to delight his mother and feed the family is now the man who sees wildlife as potentially a major source of meat and income for the rural people. He believes that wildlife should be reared and harvested like any other renewable resource. To this end, with its large and protected animal population, the Parks and Wildlife Estate is a great reservoir from which to redistribute wildlife back into its original haunts in other parts of the country.

'Commercial farmers,' observes Dr. Nduku, 'are enthusiastically turning to mixed cattle and game ranching; some have even abandoned cattle altogether. This year, the Wildlife Producers' Association want us to provide about 15,000 animals; that's a measure of the swing towards game ranching in Zimbabwe.'

But while commercial farmers need little convincing of the benefits of running game on their rangeland, extending this awareness to the peasant farmers in communal areas is another matter. It is here that Willie Nduku sees the greatest challenge – and hope – for the future.

'These are exciting times,' he says. 'The lives of a great many of our rural people, whose need for increased natural resource productivity is greatest, could improve as a result of changes in attitudes towards wildlife utilisation.'

Dr. Willie Nduku, Director of National Parks and Wildlife Management.

A Trust in Conservation

There is something of a hot-line between the desks of John Pile, Executive Director of the Zimbabwe National Conservation Trust, and Victoria Chitepo, the Minister of Natural Resources and Tourism, a few blocks apart in central Harare.

In her personal capacity, Victoria Chitepo is President of the Trust – but, in her ministerial capacity, she is always keen to know what the many active non-governmental organisations (NGOs) in Zimbabwe are thinking and doing about conservation and resource management. So much interest is shown in this area that John Pile has a very high regard for Government attitudes generally to conservation.

The Conservation Trust, which coordinates the activities of a score of member societies, is an effective bridge between the NGOs and Government. Recently, for example, it has been representing the Zambezi Society

The key to bringing the conservation message to children lies in sparking their enthusiasm and having the involvement of their teachers at school.

in its concern about the environmental impact of large-scale settlement following the eradication of tsetse flies in the Zambezi Valley.

Many special-interest groups, youth groups, private aid agencies and others – representing a wide range of conservation activities – operate throughout Zimbabwe. Their interests include trees, ornithology, herpetology, geography, archaeology, hunting, angling, falconry, traditional medicine and mountaineering. Each makes a contribution to the overall conservation ethic – both in a practical sense and in fostering the general concept of conservation. As Zimbabwe's National Conservation Strategy notes: they 'have combined with the National Conservation Trust to voice concern over conservation matters and to offer assistance with the implementation of the National Conservation Strategy. This valued public-spirited support is welcomed by Government and will be encouraged.'

Formed in 1974, the Conservation Trust also raises funds (at home and abroad) for conservation in Zimbabwe. It has helped to finance more than 150 projects – most notably the Rhino Survival Campaign, in support of the Ministry of National Parks and Wildlife Management's efforts to combat poachers in the Zambezi Valley. The money came from many sources, including companies, schools, local wellwishers and various overseas organisations. In its latest financial year, the Trust topped the million-dollar mark in money and goods for conservation.

John Pile, a veteran administrator, environmental educationalist, broadcaster and writer, has been working in the conservation field for many years. As a member of IUCN's Education Commission, he believes that education is the long-term answer to conservation in the future. He is impressed with locally-produced environmental education materials currently being used in schools.

'The key,' he says, 'to bringing the conservation message to children – who will have to live with the consequences of the decisions we are taking today – lies in sparking their enthusiasm and securing the involvement of their teachers at school. This is no easy task, because teachers already have a heavy workload and are hindered by lack of finance. But the efforts they are now making are genuine – and, in the long run, genuine effort brings success.'

Chapter Three

WILDLIFE: PAYING ITS WAY

Attitudes to Wildlife

The concept of the King's game, which goes back nearly a thousand years to the edicts of William the Conqueror, first came to Zimbabwe in 1890. With the arrival of the British colonial rule, the indigenous people suffered a traumatic double alienation. They were cut off from more than half the land in the country; and they were banned from killing wildlife – even on the land allocated to them. Suddenly, the British became the gamekeepers and the African people became the poachers. Yet, for centuries, these people's lives had been intimately, even spiritually, linked with wildlife.

In both Shona and Ndebele traditions, wild animals are seen as an integral part of the community resources, along with other natural products such as water, trees and grazing land. These are all available to the people as a whole, with regulations and religious rituals to control their use. The religious rationale holds that the true 'owners' of the land – and all its resources – were the founding ancestors. Since the ancestral spirits still watch over the land and the welfare of their children, their descendants are given certain 'rights'. The spirits exercise their continuing concern and control by speaking through mediums, who give voice to their wishes. These traditional beliefs and practices persist today and remain strong in rural societies and, to a lesser extent, in the urban centres.

Traditionally, hunting was controlled by the fundamental ethos that required the hunter to share, with the community, game collected from the communal property. After presenting a piece of the animal to the chief, who was a symbol of the ancestral spirits, the hunter shared his kill with the people. These traditions were rocked by the imposed colonial laws. When wildlife became the property of the state, hunting became 'poaching'. It persisted, of course – and still does.

But worst of all, the traditional communities were obliged to bear the costs of animal depredations while having no meaningful say in wildlife management. They came to regard it as a liability, rather than an asset. In this atmosphere of conflict, where communal farmers would rather be rid of wildlife than tolerate its presence, the conservation message had no meaning whatsoever.

In 1960 the Wildlife Conservation Act marked the start of cautious legislative reform. Large-scale (white) farmers and ranchers were given increasing freedom in making profits from wildlife on their land. Consequently, there was an immediate upsurge of interest in wildlife. By 1970, it was clear that this approach was in the best interests of wildlife on private land. Furthermore, the principles could be extended to the communal areas where the resource was 'common property'.

The result was the Parks and Wildlife Act of 1975. Still in force today, this Act is widely regarded as one of the most enlightened pieces of legislation of its kind in Africa. For the first time, the 'revolutionary' concept of transferring responsibility for wildlife management from the state to the people became a reality. However, at present it tends to be mainly owners of larger ranches who have benefited. The challenge is to bring the same benefits to people in communal lands.

Opposite:
As fish are a wild resource, conservation principles must be carefully followed to secure the necessary balance between different fish species, and to ensure that yields are sustainable from year to year.

In both Shona and Ndebele traditions, wild animals are considered to be part of a community's resources. (Woodcut by Zebedee Chikowore, Cyrene School; collection of the National Archives of Zimbabwe).

The wheel has *almost* turned full circle. In practice, the Department of National Parks and Wildlife Management remains the state authority responsible for all wildlife in communal lands – until individual communities actually show that they are sufficiently interested (and have the ability) to go it alone. But this is now happening, albeit slowly. With the increasing value of wildlife products – and the many examples of its profitable use on commercial ranches – several communal authorities have set aside land specially for wildlife management. Others have indicated an intention to follow suit. In well-populated areas, under other systems of production, some communities have even decided to relocate a proportion of their people to make way for commercial wildlife utilisation.

As human population pressures grow, people will tolerate wildlife (or encourage it) only if it provides them with tangible benefits. Consequently, the main thrust in professional conservation circles in Zimbabwe, both within and outside protected areas, is towards controlled, sustainable utilisation of wildlife. This includes consumptive uses, such as sports hunting and cropping, as well as non-consumptive use for tourism. Since the biggest threat to wildlife as a whole is habitat destruction, this realistic approach is intended to provide an 'umbrella' under which the entire spectrum of other species should be able to survive.

On Safari

When a safari hunter squeezes the trigger on his sporting rifle – and a magnificent wild animal is brought down – it is certain to upset many people. But these days sentiment has little to do with *true* conservation. The modern view is that wildlife must be nurtured as a usable natural resource if conservation practices are to be taken seriously.

As in many African countries, Zimbabwe's safari hunting has become a controlled, structured and profitable way of using wildlife as a renewable natural resource. It involves a relatively small offtake of trophy

Wildlife utilisation is often the most efficient way of using poor quality land. Processing the skins and ivory within Zimbabwe can provide much revenue and employment. However, strict controls on the wildlife industry are essential in order to avoid abuse of the resources. (ZTDC)

animals, with a far greater revenue per animal than is possible when cropping for meat. With careful management, it clearly makes both economic and ecological sense. It puts cash into community development projects in remote parts of the country; and it puts much-needed foreign exchange into the Treasury in Harare.

Commercial safari hunting is a relatively recent development in Zimbabwe. Only 30 years ago, sports hunting was confined to private land. Then the 1960 Wildlife Conservation Act provided for the establishment of Controlled Hunting Areas, the first of which was opened, in the Zambezi Valley, in 1961. Demand for space in the hunting camps was so enormous that, by the late 1960s, the Government began to take an interest in commercial wildlife utilisation.

A depressed cattle-ranching area, in the north-west of the country, was bought and handed over to the Department of National Parks and Wildlife Management for safari hunting. This was divided into seven concessions which were leased out in 1973 and became the Matetsi Safari Area. The Matetsi experiment was a springboard for the development of safari hunting as a legitimate form of land use – and soon encouraged the leasing of more concessions to commercial safari operators.

Now Zimbabwe is taking an innovative step farther along the safari trail. It is making a concerted (and perhaps unique) effort to develop commercial hunting and other forms of wildlife utilisation in some of the most marginal parts of the country. In this way, areas which are not otherwise commercially exploitable are now being made productive.

Wild land for hunting in Zimbabwe now amounts to about 65,000 square kilometres – an impressive seventeen per cent of the country. It is included in protected areas, forests, commercial farms and communal farmland. Hunting in these areas is permitted throughout the year – but occurs mainly during the cool dry season, from April to August, and the hot dry season, from September to October or mid-November.

Quotas for trophy animals in each safari area and communal land concession are set by the Department of National Parks and Wildlife Management. They are calculated on the basis of aerial censuses and ground

Trophy hunting is a multi-million dollar business in Zimbabwe; but the Government is careful to sustain the wild resources upon which it depends. Eight per cent a year is the maximum rate of offtake from a population of big cats.

reconnaissance by their staff, and using reports from professional hunters. Quotas range from 0.5 per cent of the estimated population in a given area, as in the case of elephants, to two per cent for the larger ungulates such as buffalo, sable, waterbuck, eland and zebra, to eight per cent for large cats.

Hunting concessions in the communal lands are also administered by the Department of National Parks and Wildlife Management, even though the land itself comes under the jurisdiction of the Ministry of Local Government and the District Councils. Lease and trophy fees are collected by the Government. They are then returned to the District Councils in the form of grants for approved development projects.

Until recently, communal land concessions were remote and largely uninhabited areas, often adjacent to Parks and Wildlife land. But expanding human populations – and immigration following the eradication of tsetse flies in the Zambezi Valley – have led to a rapid increase in subsistence farming and an influx of livestock. Such developments now threaten the viability and continuity of wildlife utilisation in these areas. The number of hunting

Zebra graze alongside a herd of crossbred Hereford-Brahmans at Imire Game Park near Marondera. The Zebra graze off the sour grasses in the vlei, making way for the more palatable growth which is preferred by cattle. Antelope of various species make use of tree browse, which is not used by livestock. According to owners Norman and Gill Travers, this full use of the open woodland habitat means that 1500 cattle and 600 head of game can run where once only 1500 cattle were stocked.

Norman Travers put up the first game fence on his beef-and-tobacco farm in 1972. Since then, with animals translocated from wild areas by the Department of National Parks and Wildlife Management, Imire has built a first class game-viewing operation. Whilst tourists provide a good source of revenue – backed up by other operations such as cropping for venison and now safari hunting with bows and arrows – Mr. Travers considers conservation to be Imire's main objective: 'Many species are thriving here, under perfectly controlled conditions'.

concessions in the communal lands is dropping. The need for appropriate institutional mechanisms and innovative programmes to promote the involvement of communal farmers in wildlife management is therefore becoming increasingly urgent. This is one of the greatest challenges facing this field of development.

It is on commercial farmland that hunting is developing fastest. Under the Parks and Wildlife Act of 1975, landowners or occupiers are responsible for the management and utilisation of wildlife on their property. This gives commercial farmers the latitude and incentive to manage their wildlife resources on a sustainable and profitable basis. Consequently, many extensive cattle ranches have either incorporated wildlife, or have switched entirely from cattle to wildlife systems which depend on commercial hunting and safari operations for their profits. About 30,000 square kilometres of ranchland in the commercial farming sector is currently being used for game ranching and safaris. The potential for further development is enormous – in as much as two-thirds of Zimbabwe – even if only for antelope, gamebird and waterfowl hunting rather than for big game. Furthermore, safari operators and commercial farmers have not yet explored the potential for joint wildlife ventures with communal area farming communities.

The commercial safari industry is becoming a significant foreign currency earner for Zimbabwe. The outlook for substantially increasing this revenue is good, given the infrastructure already in place, the expertise of the operators, the support of Government, and the very significant possibilities for developing hunting operations in the communal lands.

Game Ranching and Farming

On a global scale, wildlife is a scarce commodity. Unlike many agricultural products, it is unlikely to face serious competition in world markets, where surpluses mean poor prices to producers in developing countries.

In Africa, Zimbabwe has, for a long time, been a leader in the conservation and management of this commodity. But only now is it learning to exploit wildlife to the full.

Legislation enacted in 1975, enabling landowners to make use of the formerly state-owned wildlife on their land, was the springboard for what is now a thriving and rapidly growing industry. The movement is so strong that game ranching in Zimbabwe could be on the brink of the same dramatic transformation that occurred in agriculture 30 years ago.

The Wildlife Producers' Association, formed in 1985 by the powerful Commercial Farmers' Union, now has about 500 rancher and farmer members who have either switched entirely from livestock to game animals, or are running wildlife in addition to cattle.

It makes good ecological and economic sense. Low rainfall and poor soils mean that about two-thirds of Zimbabwe is suitable only for animal production, with some dryland cropping with drought-resistant crops. In these regions a major cause of erosion, soil loss and declining fertility is overgrazing by the domestic animals. But risks of overgrazing are less – and inputs lower – if a combination of livestock and wildlife is used in these arid areas. Cattle do not have the capability to make full use of the habitats available – while indigenous large herbivores produce as much (if not more) protein per hectare than domestic livestock under the same conditions. In commercial terms, systems which incorporate both wild and domestic stock together are often more profitable than either on its own.

Game *ranching* takes place on very large properties. These are mainly in areas where game animals have always been plentiful, though the ranchers also buy in animals to increase the stocks of particular species. In addition to hunting safaris, the ranchers crop animals to sell the venison and hides. Those with surpluses of desirable species sell them to other ranchers and farmers.

Game *farming*, which is combined with both cattle and crops, takes place mainly in the better rainfall highveld areas. These mixed farms are generally much smaller than the ranches and closer to urban centres. Consequently, they attract visitors for short periods of game viewing on foot and horseback – which provides an extra source of income. Most of the game animals, which are kept in large fenced paddocks, have to be introduced.

Both ranchers and farmers rely heavily on the Department of National Parks and Wildlife Management for their stock. The animals are supplied by capture units in a system (involving the Wildlife Producers' Association) supported financially by a revolving fund administered by the Zimbabwe National Conservation Trust. It works well – but demand by far exceeds supply. So private capture units are also now operating, to help meet the shortfall in Zimbabwe and to fill overseas orders.

The animals most in demand are zebra, eland, impala, sable, waterbuck, tsessebe, wildebeeste and warthog. But specialist wildlife production is also well established and has considerable further potential. There are substantial crocodile and ostrich farming enterprises – and even python farming is a possibility.

In addition, the Wildlife Producers' Association contributes directly to the conservation of endangered species. Members have received breeding nuclei of black rhinoceros, Lichtenstein's hartebeeste and roan antelope, with the long-term objective of re-stocking their original natural haunts.

A Study for Africa

A major land-use project underway in Zimbabwe could have far-reaching benefits for Africa. The five-year exercise, launched in mid-1988, is designed to test and develop multi-species systems of wildlife as an ecologically and economically sustainable option for land use on the continent.

Funded by the World Wide Fund for Nature (WWF), it is hoped that the project will act as a catalyst in a variety of research programmes aimed at improving management in the African savannas. It will collaborate with, and critically study, a range of pilot schemes in different parts of Zimbabwe, including privately-owned commercial farmland and subsistence farming areas under communal ownership.

The project is examining many aspects of wildlife production systems – alone and in combination with livestock. In particular, it will look at the influence of these systems on species diversity, nutrient cycling, and ecosystem stability and resilience. The aim is to test a series of hypotheses, about managing land for wildlife and livestock, formulated for African savannas with an annual rainfall of less than 800 mm. These suggest that multi-species systems of land use are:

- financially more productive than extensive single domestic livestock systems, under both private and communal ownership;
- able to support a greater diversity of plant and animal species (excluding large mammals) than single domestic livestock systems;
- both more resilient and more stable than single species systems in drought-prone environments;
- economically and ecologically more sustainable in low rainfall areas and require lower subsidies than extensive cattle ranching.

This is an important project that will strengthen the existing involvement of the Zimbabwean Government and non-governmental organisations in multi-species production programmes.

A Pharmacy from the Wild

Before the missionaries and colonial settlers brought scientific medical practices to the region, traditional healers were the doctors in Zimbabwe – treating their patients with medicines prepared from trees, shrubs, herbs, roots and animal matter. Today, despite successful Government efforts to extend modern health services throughout the country, they continue to play a central role in society.

Traditional healers and modern doctors have more in common (and more to learn from each other) than many people realise. Against the history of suspicion, distrust and hostility that has existed between them for many years, they are beginning to look at each other with renewed interest – and even some respect.

The main difference between their approaches is that scientific medicine is concerned with physiological disturbances in the body, whereas traditional medicine is more concerned with psychological and social problems. While the two are unlikely ever to meet, in Zimbabwe they are at least making some effort to understand each other.

Nowadays, traditional medicine is actually encouraged rather than merely tolerated. There is no anomaly in this. White-coated doctors, trained at the excellent University of Zimbabwe Medical School, and skin-clad traditional healers, with their potions, are both treating the sick in their own way. Their coexistence is both pragmatic and sensible in a country where there are still not enough doctors and where, in many rural areas, clinics and hospitals giving formal treatment are few and far between.

But the fact that traditional medicine is also flourishing in the towns and cities is indicative of its continuing strong appeal, even in modern society. Many people still prefer to visit the traditional healer rather than the doctor's surgery or the chemist. There are two main reasons for this: the undoubted efficacy of many traditional medicinal preparations; and, perhaps more important, the enduring belief that bodily ailments can be the manifestation of spiritual forces – which only traditional healers are considered able to deal with.

It is thought that between 60 and 70 per cent of all patients who seek medical help in Zimbabwe go at some stage to a traditional healer.

In urban areas, a patient's choice as to which practitioner to consult – the traditional or the scientific – may be a difficult one. The most important determinant is whether he considers his ailment to be the result of normal or abnormal causes. If it is regarded as normal he is as likely to see a clinic or doctor as he is a traditional healer; sometimes, to be on the safe side, he will do both. If, however, he regards the ailment as abnormally inspired, he will almost certainly go to the traditional healer. But complications arise because almost any illness may be regarded as normal at one time and abnormal at another.

Some quackery certainly exists among the traditional healers. But in Zimbabwe there is a controlling influence, in the form of the Zimbabwe National Traditional Healers' Association (Zinatha). Formed in 1980, Zinatha and its disciplinary council (which was created by the Traditional Medical Practitioners' Act) holds a register of the healers, regulates their work and encourages further research.

Zinatha's President, Professor Gordon Chavunduka, Dean of Social Studies at the University of Zimbabwe, firmly believes that traditional medicine will be as relevant in Zimbabwe in the 21st century as ever before. He sees its holistic approach to the treatment of disorders – embracing the psychological, social, cultural and spiritual dimensions – as playing an important role in everyday life.

He also sees the development of a local pharmaceutical industry, based on tried and tested traditional remedies prepared from natural products. This would not only save foreign currency (through import substitution) but would also earn it through exports. Zinatha has received several enquiries from overseas about traditional medicines produced here. Furthermore, Zinatha is beginning to have a positive influence in conserving medicinal plants. For example, it is cultivating wild species in a plot allocated by the Harare Municipality.

Above, top to bottom:

Wounds on the bark of Ozoroa insignis *trees are a frequent sight – traditional healers say they obtain an aphrodisiac from the bark.*

The tuber of Dicoma anomala *is used as an infusion for dehydration and for depressed fontanel in babies.*

Strychnos spinosa *is used as an emetic (a medicine to cause vomiting) when unripe, but it is also a refreshing fruit once ripened.*

A priest's son, Gordon Chavunduka spent his boyhood at St Augustine's – a large mission near the town of Mutare on the Mozambique border. The mission hospital provided for the family's needs, so he had no contact whatsoever with the traditional healers who attended to so many others living in the area.

But at university he was influenced by the pioneering research into traditional medicine being carried out by the late Professor Michael Gelfand, a world authority on the subject. Professor Chavunduka now undertakes his own research, and is a frequent lecturer to doctors and nurses, concentrating on the social aspects of traditional medicine.

Progress is, perhaps inevitably, a little slow – the Professor acknowledges that there is scepticism and opposition. But he reports too an open-mindedness and interest in the subject which did not exist even a few years ago.

*Most crops which are cultivated in Zimbabwe came originally from other countries. Yet Zimbabwe itself has a tremendous range of wild foods, which between them can provide the full complement of nutritional needs. In future, some of these foods could become increasingly important in drought-prone areas. From left to right 'fruit of the hissing tree', false medlar, baobab pods, elephant love fruit, an edible fungus (*Cantharellus miniatescens*), a caterpillar of the msasa moth.*

(MCP, DH, DR, MCP, DH, DH)

Food from the Wild

Zimbabwe has a rich and varied natural food supply. Fruit, mushrooms, caterpillars and winged termites, among others, all feature in the Zimbabwean diet – gathered for the pot by village people and obtained from urban markets by a great many town and city dwellers who have not lost their taste for the traditional.

There is fruit for the picking. Much of the country is covered with open woodland and many of the trees and shrubs produce edible fruit. Some trees are protected by custom, and left standing in cultivated lands, while seedlings often germinate along well-used paths and around villages.

No fewer than 20 species of figtree are indigenous to Zimbabwe – and all produce edible fruit, though some are less palatable than others. The sycamore fig *(Ficus sycamorus)* produces ripe figs throughout the year – on different trees, thanks to their intricate relationship with a small pollinating wasp.

Many wild fruits, such as those from the waterberry *(Syzygium guineense)*, are watery and help quench the thirst. These are generally eaten on the spot, rather than harvested. But when no such refreshing fruit is available, the inner bark of the ubiquitous *Brachystegia boehmii* is easily stripped off and chewed; as a bonus, the fibres become pliable and form a widely used and surprisingly durable rope.

The larger, sweeter fruits are collected for use in the villages or for transporting to market. Notable among them is the widespread mobola plum *(Parinari curatellifolia)*. It is a peculiar tree because periodically (and inexplicably) it releases an odour that is reminiscent of an overflowing sewer. However, its prized oval fruit, the size of a plum, has a yellow flesh that is sweet and tasty. Since it is an evergreen, *Parinari* also provides dense shade – most welcome in the dry heat just before the rains.

At medium altitudes, the stony slopes that are poorly suited to cultivation often support stands of the mahobohobo *(Uapaca kirkiana)*.

The females of this tree bear a heavy crop of round fruit about three centimetres in diameter. These begin to ripen during the late dry season, when the popular sweet flesh is sold along roadsides and in local markets.

Strychnos cocculoides and *Strychnos spinosa* are small trees that become decorated with a hard green fruit the size of a large orange. When the woody shell turns yellow, it is cracked open and the flesh, which liquefies and begins to ferment, is delicious.

The marula tree *(Sclerocarya birrea)* is common at medium to low altitudes and produces a very sweet and tasty fruit. Its vitamin C content is reputedly four times higher than that of orange juice. The fruit ferments easily and is also made into an alcoholic drink.

The few wild fruits that keep well are transported – often long distances – to urban markets. Principal among them is *Ziziphus mauritiana*, a grape-sized fruit with a sweet, well-flavoured flesh. It partially dries out and so stores well.

Ziziphus fruit is most prolific at lower altitudes, in the hot dry season. Huge numbers are brought into local markets, where their rich honey scent fills the air around the trading stalls. Although often eaten raw, the fruit also forms the basis of 'kachaso' – a potent drink that is distilled illegally.

The fat and ancient baobab tree *(Adansonia digitata)* bears fruit with a hard casing around a dry, floury pulp in which the seeds are embedded. This pulp has a high concentration of tartaric acid and potassium bitartrate and is refreshing to suck. A tasty drink is also made by soaking the seeds in water. Even the leaves have a strong flavour; they contain a mucilage that binds the vegetable and meat relish served with sadza, the stiff maize meal porridge that is the staple dish of most Zimbabweans. Other plants with a high mucilage content are also used, including leaves from the cultivated *Hibiscus* and the fruit okra *(Abelmoschus esculentus)*.

Another tree in the hibiscus family is the indigenous *Azanza garckeana* whose fruit – the indelicately named snot-apple – contains a highly prized sweet slime. The Ndebele people, whose language is richly embroidered with click sounds made with the tongue on the roof of the mouth, call the fruit *uxhakhuxhaku*. With its multiple clicks, this aptly describes the sound made while chewing on the glutinous slime.

Most of the edible berries in Zimbabwe are small. Although profuse, they are not harvested but generally eaten straight off the bush. They provide thirst-quenching snacks (and valuable sugar and vitamins) for children herding cattle or walking to and from school. There are a great many of these berries, including some sticky ones in the genus *Grewia* and the brown chocolate-like ones of *Vitex payos*, that stick to the roof of the mouth.

A variety of palatable mushrooms are gathered, for subsistence or sale. But cases of illness and death are not uncommon and, because of the risk of eating poisonous types, many people leave all wild mushrooms alone.

Two species of caterpillar, associated with particular host trees, are widely collected from the branches of the mopani *(Colophospermum mopane)* and mukarati *(Burkea africana)* respectively. The gut contents are squeezed out before the caterpillars are boiled in water, or dried in the sun or over a fire. Very high in protein, they keep well and are sought-after at local markets. They are eaten dry and crunchy, or lightly fried as part of a relish.

Finally, there are winged termites, or flying ants as they are commonly called. High in oils and very palatable, they are very popular. Their flight is triggered by humidity and the sound of rain – and can be induced by beating on the termite mounds. The hapless insects are then trapped as they emerge in profusion.

Fishing

Fish is a growing part of the Zimbabwean diet. Both government and university research is determined to make this valuable, versatile and renewable food resource increasingly abundant and more widely available.

But unlike neighbouring Zambia, Zimbabwe has no natural lakes and relies almost entirely on dams for commercial and subsistence fishing, with some small-scale activity in the river systems. Total fish production is between 20,000 and 30,000 tonnes a year – half of this from commercial fisheries.

Above:
Although Zimbabwe is home to 122 indigenous fish species, it is the tiny kapenta, introduced to Lake Kariba, which makes up 93 per cent of the national catch.

Opposite:
Zimbabwe has no natural lakes, and relies almost entirely on dams for its fishing industry.

Zimbabwe has a fairly rich fish fauna, with 122 indigenous species recorded. This diversity stems largely from the country's connections with the Zambezi river system. With about 340 known species (including those of Lake Malawi) the system ranks equal with the Nile and second only to the Congo, which supports some 600 different fish.

At least 30 species have been introduced to Zimbabwe – for angling, for aquaculture, or to increase production in dams. The most spectacular and successful of these has been the Tanganyika sardine *(Limnothrissa miodon)*, known locally as kapenta. It was introduced by Zambia from Lake Tanganyika in 1966-67, to fill the vacant deep-water ecological niche in the young Lake Kariba. But it quickly colonised the entire lake and now supports a multi-million dollar fishing industry from both shores, providing employment and supplying cheap protein to rural and urban people. Operating at night, the kapenta rigs dipnet the shoals of small fish as they rise to the lure of powerful lights. Most of the catch is brined and sun-dried; the rest is frozen, canned or used as crocodile food.

Ninety per cent of the total fish production in Zimbabwe originates from Lake Kariba. Fish are an abundant (but finite) resource in the lake, which is still in the process of adjustment from a riverine to a lacustrine ecology – following the damming of the Zambezi River in 1959. It is now classified as 'mesotrophic', i.e. of moderate nutrient status. In the early years of impoundment, the inundation of so large an area of vegetation resulted in highly nutrient-rich water which yielded exceptional catches of fish. Now that the lake is settling down it is, like the Zambezi itself, less full of nutrients. Nevertheless, it still produces about 10,000 tonnes of fish a year – of which 93 per cent is kapenta.

Gillnet inshore fishing, along the shoreline of Kariba and other sizeable waterbodies (such as Lakes Kyle, Robertson, McIlwaine and Bangala) yields good catches of palatable white fish. These are mainly indigenous and introduced species of bream, or Tilapia.

Farming of Tilapia is increasing, with several experimental schemes and a number of commercial producers. Significantly, small-scale commercial farmers are also seeing its value (for both revenue and food) and are being encouraged in their efforts to utilise the fish.

In the cold Eastern Highlands, trout is also farmed. The fingerlings are bred by the Department of National Parks and Wildlife Management for sale to commercial operators who grow, crop and market them for the table.

Zimbabwe's fish farming, or aquaculture, probably yields not more than 1,000 tonnes of fish a year. But it is still in its infancy – and already contributes in socio-economic terms by providing jobs, saving foreign currency on imports, and supplying hotels, restaurants and shops which have difficulty in getting seafood.

There is considerable potential in fish farming and, with current government efforts to promote fishing cooperatives for net fishing, the prospects for continued growth are good.

Chapter Four

A NATION OF FARMERS

Feeding the Nation

The importance of agriculture to Zimbabwe cannot be over-emphasised. It plays a locomotive role in the economy. It feeds the nation, with surplus for export. It employs 27 per cent of the working population. It contributes over fifteen per cent to the Gross Domestic Product. It generates 40 per cent of the country's foreign exchange earnings. And it provides the bulk of raw materials required by the manufacturing industries. Zimbabwe's agricultural prowess is also the kingpin in the Southern African Development Coordination Conference (SADCC) regional food security programme. Its performance is impressive by any standard.

Farming methods range from modern tractors and combine harvesters to ox-drawn ploughs and the hoe. But production patterns continue to be dictated by historical developments relating to the division of land since the turn of the century. This has resulted in four distinct types of farming area:

Large-scale commercial farms
Formerly set aside as white-owned farming land, these areas make up about 80 per cent of ecological regions 1 and 2 (see *A Capacity to Grow Food*) and some 40 per cent of the total land area of Zimbabwe. The number of commercial farms has decreased since independence, partly because of the Government's policy of land distribution through resettlement programmes. But there are about 5,500 of them remaining (with an average size of 2,200 hectares) and commercial farming still produces most of the country's market surplus.

Small-scale commercial farms
Formerly called African Purchase Lands, these areas make up four per cent of the total land area of the country, with about 75 per cent of the farms being in regions 3 and 4. The number of small-scale commercial farms – about 8,600 – has remained constant over the past decade. They have an average size of 124 hectares.

Resettlement areas
These have been established since independence, as part of the Government's land redistribution efforts. The Government buys commercial farming land and makes it available to farmers from those communal lands which are under pressure from people and their livestock. The average farm in resettlement areas is five hectares, with the additional use of communal grazing land.

Communal land areas
Formerly known as Tribal Trust Lands, these areas make up 42 per cent of the country and are home to about 4.3 million people. Almost 75 per cent of the farms are in the least productive, ecologically fragile regions 4 and 5. The average farm size is about 23 hectares.

Zimbabwe's main food crop is white maize, the staple food of most of its people and grown in all four farming areas. Except in drought years, when yields are low and local sales high, maize stocks held by the Grain Marketing

Opposite:
Zimbabwe's cotton is of excellent quality, and forms one of the biggest exports in both lint and textile form.

Agriculture employs 27 per cent of the wage-earning population, and provides extra work at harvest time. From left to right: harvesting grass, collecting chickens' eggs, gathering leaf vegetables, plucking tea, winnowing soya beans and picking cotton.

Board allow for sizeable exports. A wide range of other food crops is grown, including three more summer grains: sorghum, mhunga and rapoko (which is known elsewhere in the world as millet); irrigated winter cereals, notably wheat, barley and oats; soyabeans, groundnuts and sunflowers; vegetables; and citrus and deciduous fruit.

The main cash crops are tobacco, cotton, sugar, tea and coffee. The flue-cured tobacco industry is crucial to Zimbabwe's economy and its people. It is a major contributor to Gross Domestic Product and accounts for no less than 20 per cent of the country's foreign exchange earnings. The industry provides a livelihood, directly or indirectly, for over half a million people. Twelve out of every 100 workers in the country are employed in the tobacco industry, and 33 out of every 100 farm workers are employed in tobacco farming. Zimbabwe's 'leaf of gold' enjoys an international reputation for quality. It is sold by auction in Harare – in what is said to be the largest tobacco auction floor in the world. The tobacco is exported to more than 60 countries, with about half the sales going to the European Economic Community.

The cotton industry continues to grow steadily. There are now more than 190,000 registered farmers in Zimbabwe, compared with 30,000 in 1980. A crop of 460,000 tonnes is forecast by 1991. Cotton supplies a major ginning, spinning and textile industry; it is also exported in lint and cloth form.

In the hot and dry south-eastern lowveld, about 40,000 hectares of sugar cane thrives under flood and sprinkler irrigation. They yield a peak of nearly half a million tonnes of raw sugar, for the local market and for export. Valuable by-products from the milling and refining process include molasses, for industrial and potable alcohol and for cattle feed. A sizeable percentage of the cane is converted to ethanol for blending with petrol. The ethanol plant was a first for Africa.

Zimbabwe's annual tea production of about 60,000 tonnes represents no more than two or three days' world consumption. Nevertheless,

Sugar is one of Zimbabwe's main cash crops. A sizeable percentage of the cane is converted to ethanol for blending with petrol. The ethanol plant was a first for Africa. As much as half a million tonnes of raw sugar is produced every year, for the local market and for export.

the yield is exceptionally high, the quality is good, the industry is efficient, and the prospects for export growth are promising. About 40 per cent of the total tea-growing area is now under irrigation, thus significantly increasing yields.

High-quality green coffee from Zimbabwe is much in demand overseas, where it is used mainly for blending. Production could expand dramatically – but the present marketing constraint is an unrealistic share of the quality world market, which is controlled by the International Coffee Agreement.

Horticulture has become one of Zimbabwe's fastest growing industries. Cut flowers, cultivated under controlled conditions, are air-freighted to the winter market in Europe, where they are well received at auction. Tropical and sub-tropical fruits also realise good prices. Given adequate air transport, there is considerable potential for horticulture to become a major export earner.

There are a number of significant commercial livestock activities, including pigs, sheep, goats and poultry. But most of all, Zimbabwe is beef country. However, considerable expertise in breeding and management results in a high-quality product so popular in Europe that the size of current exports makes it impossible to satisfy the growing home market. There are 4.83 million head of cattle in the country – 65 per cent on the communal lands and the remaining 35 per cent in the commercial farming areas.

The commercial herd (which supplies 90 per cent of the beef for the local market and all of the export beef) has declined dramatically from its 2.83 million peak in 1977. This is the combined result of increased hostilities at the end of the civil war, three droughts and a diminishing confidence in the industry caused by eroding returns. In contrast, dairy production from the communal herd (which is kept mainly for traditional, rather than economic, reasons) is booming. Investment since 1980 has been substantial and Zimbabwe is the only country in the SADCC and PTA regions with a viable dairy industry with export capacity, plagued though it is from time to time with packaging shortages.

Right and below right:
Two of the tremendous advances since Independence are increased maize harvests and wealthier small-scale farmers. In 1988, 66.8 per cent of the maize crop was produced by communal area farmers, compared to only 4.8 per cent in 1979.

Increasing input from communal land peasant producers entering the cash economy, and small-scale farmers, has been an outstanding feature of Zimbabwe's overall agricultural development since independence. This reflects their access to extended credit facilities, their readiness to diversify and adopt new techniques, and their ability to buy fertiliser and equipment. In the 1988 season, 66.8 per cent of the national maize crop was produced by communal area farmers. This is a great improvement on the 1979 figure of 4.8 per cent.

It is a trend which holds out promise for the overall development and raising of productivity and living standards in the communal areas. But the prospects for the future depend upon careful conservation of all the natural resources which form the backbone of the Zimbabwean agricultural industry – in particular, soil, water and the wide genetic base of both crops and livestock.

Opposite:
Zimbabwe's coffee routinely fetches some of the highest prices on the world market.

A Capacity to Grow Food

Despite Zimbabwe's considerable capacity to feed itself, less than 20 per cent of the country is ecologically suited to intensive crop production. In all, at least three million hectares are under the plough; but considerable cultivation takes place in areas that are either unsuitable for the crops being grown or unsuitable for cultivation at all. These areas are suffering, in varying degrees, the economic and ecological consequences of accelerating soil erosion.

To understand why four-fifths of Zimbabwe is unsuited to intensive agriculture, it is necessary to recognise the country's five 'agro-ecological regions', each distinguished by soil type and climatic factors.

Region 1: Specialised and diversified farming
Covers an area of only 7,000 square kilometres – less than two per cent of the land area of the country. Annual rainfall is over 1,000 mm in areas below 1,700 metres, and over 900 mm at higher altitudes. Agricultural activities are mainly forestry, fruit, intensive livestock and, in frost-free areas, tea, coffee, macadamia nuts and other plantation crops. About 74 per cent of farming is on large-scale commercial land, 24 per cent on communal land and two per cent on small-scale commercial land.

Region 2: Intensive farming
Covers an area of about 58,600 square kilometres – fifteen per cent of the land area of the country. Annual rainfall is from 750 mm to 1,000 mm. Agricultural activity is centred around intensive crop and livestock production. Crop yields are affected in some places by relatively short rainy seasons or dry spells during the season. Almost 74 per cent of farming is on large-scale commercial land, 22 per cent on communal land and four per cent on small-scale commercial land.

Region 3: Semi-intensive farming
Covers an area of about 72,900 square kilometres – nineteen per cent of the land area of the country. Annual rainfall is from 650 mm to 800 mm, with fairly severe mid-season dry spells. Farming activities are mainly livestock and fodder and cash crops, with marginal production of maize, tobacco and cotton. About 49 per cent of farming is carried out on large-scale commercial land, 43 per cent on communal land and eight per cent on small-scale commercial land.

Region 4: Semi-extensive farming
Covers an area of 147,800 square kilometres – almost 38 per cent of the land area of the country. Annual rainfall is from 540 mm to 650 mm, with periodic seasonal drought and severe dry spells during the season. Farming activities are primarily livestock and drought-resistant crops. Communal land carries 62 per cent of the farming, large-scale commercial land 34 per cent, and small-scale commercial land four per cent.

Region 5: Extensive farming
Covers an area of 104,400 square kilometres – 27 per cent of the land area of the country. Annual rainfall is too low and erratic for even drought-resistant fodder and grain crops. Farming activities comprise mainly cattle and game ranching. About 45 per cent of farming is carried out on communal land, 35 per cent on large-scale commercial land and under 20 per cent in National Parks.

Above:
Over the years, food habits change and bread is now as popular as sadza – or maize porridge – especially as it does not require cooking in the home. However, most wheat still has to be produced on irrigated land. Research into rain-fed wheat strains, and minimum tillage systems which do not dry the soil out, should help to satisfy the nation's increasing appetite for inexpensive bread.

Opposite, top:
There are 5,500 large-scale commercial farms in Zimbabwe, covering about 12 million hectares of the best land. Their excellent soil and water conservation systems have been at the root of their success. As pressures to produce higher yields build up, the effect of inputs such as chemical fertilizers and pesticides on the environment should be carefully monitored and action taken where necessary. (PB)

Opposite, bottom:
Sorghum is an excellent rain-fed crop, more suited to arid areas than maize.

Zimbabwe's agricultural bounty contributes over fifteen per cent to the GDP and provides 40 per cent of all foreign exchange earnings. From left to right: turnips and radishes, coffee beans, tomatoes, courgettes, chillis, maize cobs, oranges and apples, cabbages and pumpkins.

Food Security for the Region

Zimbabwe is acknowledged as the agricultural leader in the nine-nation Southern African Development Coordination Conference (SADCC) grouping, which comprises Angola, Botswana, Lesotho, Malawi, Mozambique, Swaziland, Tanzania, Zambia and Zimbabwe. Each member country has a special responsibility to the group, Zimbabwe's being food security.

The objectives of the SADCC food security programme are to meet the basic need for food; to provide a buoyant farming sector;

Pesticides help to avoid crop spoilage, and so can save Zimbabwe millions of dollars a year. But they can be lethally toxic to people, and safety clothing is essential when applying them. Less obvious, however, is their danger to the environment (building up over the years in soil, water and animal bodies). More work needs to be done to monitor the effects of agricultural chemicals and control their use.

to reduce the present heavy drain on foreign exchange imposed by basic food imports; and to reduce dependence on South Africa as a food supplier.

Among the twelve regional food security projects are an early warning system, an inventory of the regional agricultural seed production and supply, a food marketing infrastructure and improved regional irrigation management.

Zimbabwe also has responsibility for coordinating SADCC agricultural research, livestock production and disease control, fisheries, forestry, wildlife, soil and water conservation, and land use.

Top:
In many areas of Zimbabwe, soil is being lost at a rate of more than 40 tonnes per hectare per year. The problem of soil erosion is particularly widespread in the communal lands.

Above:
Soil conservation is not an entirely new concept in Zimbabwe. In mountainous parts, traces can be seen of ancient terracing. In some areas subject to heavy rainfall – such as here at Chimanimani – terracing remains an essential part of farming practice.

The Growing Problems of Soil Erosion

A two-millimetre droplet of rainwater hits the ground with a velocity of about four metres per second, or fifteen kilometres an hour. On impact its kinetic energy is expended in breaking down the soil particles. Multiplied billions of times on an exposed field, this tiny pounding becomes a mighty hammering – and causes erosion. Soil is carried away in suspension in the water. The rainwater progressively strips off the precious (and sometimes very shallow) topsoil, washes away nutrients and gouges out gullies – which, each year, become deeper, wider and longer. Carried down hillsides and into water courses, the soil particles then build up to choke rivers and silt up dams.

In parts of Zimbabwe, the problem is so severe that we are literally running out of soil. In many areas, soil is being lost at a rate of more than 40 tonnes per hectare per year. On soils unprotected by thick vegetation, and on long, uninterrupted slopes the problem is even worse. It is these vulnerable and ecologically fragile areas, comprising mainly sandy soils in low or erratic rainfall regions, which can least afford to lose this life-supporting and irreplaceable resource.

Accelerated soil erosion, due to growing pressures on the land from human and livestock populations, is particularly widespread in the communal lands. Indeed, in many cases, erosion is threatening to undermine major rural development programmes. In some areas, within just ten years, soil depth may become insufficient to grow the staple maize crop. Small dams, fed by large catchments which are subject to severe erosion, can silt up entirely in only five years. Even major dam sites are vulnerable in the longer term – putting domestic water supplies at risk and frustrating efforts to extend irrigated cropping. And, on a national scale, the value of nutrient loss in eroded soil, related to the cost equivalent of fertilizers, amounts to millions of dollars a year.

The erosion problem in Zimbabwe has its roots in the early colonial division of the country on a racial basis – and is the result of several decades of political, economic and environmental conflict. The communal lands are largely in areas of poor soil and hilly terrain. Over the years, the occupants have placed increasing (and often intolerable) strains on their land, denuding it of trees and grass cover, exhausting its fertility and exposing it to the relentless forces of erosion. The consequences of this largely enforced mismanagement are now disturbingly evident. And the problem is growing.

Soil conservation work has, however, long been a feature of development efforts in the communal areas, albeit with varying degrees of success. During the independence war, when the communal lands were the scene of considerable military activity, this work came to a standstill. But it is now pursued vigorously by Agritex, the Department of Agricultural, Technical and Extension Services of the Ministry of Lands, Agriculture and Rural Resettlement.

However, the soil erosion threat is by no means confined to the communal lands. In the last several years, soil erosion and land degradation in general has been increasing in some commercial farms and marginal lands.

Serious though the soil erosion problem is, Zimbabwe is actually better placed than most African countries to combat it. In particular, it is fortunate in having the results of excellent 1:25,000 scale panchromatic aerial photography, carried out on a systematic basis at five- to eight-year intervals since the early 1960s. A recent national soil erosion survey, which was based on the study of about 8,500 of these photographs, has provided detailed information on the extent and pattern of land degradation.

The survey shows negligible soil loss over 60 per cent of the country, particularly on commercial farmland and in national parks. More extensive – but still localised – erosion occurs over 25 per cent of the country, within both commercial and communal farming areas. But the most serious erosion, affecting more than ten per cent of the country, occurs almost entirely on communal lands.

The study is now being put to good use by policy-makers and experts on the ground to halt – and hopefully reverse – the current trends in soil erosion. But, as always, it is a fight against time.

The national cattle population is just over half the human population, at 4.8 million. Sixty-five per cent of the cattle population is farmed in communal areas, the remainder on commercial beef and dairy farms.

Men and Cattle

In Africa, man and his cattle are not easily parted. Zimbabwe is no exception. Here, cattle represent much more than meat and cash on the hoof for the traditional pastoralist. They buy his bride, pull his plough and cart, manure his fields, give him security, and are important elements in his social and religious life. In this society, a poor man is often a man without cattle.

Although the herd in the communal farming lands makes up 65 per cent of the total number of cattle in the country, it is not considered in the commercial sense to be a beef herd at all. The 'formal' slaughter-for-cash offtake from it is only about two animals in a hundred. In contrast, the offtake from the smaller herd on the commercial farming land is as high as 23 animals in a hundred. Indeed, the commercial herd supplies 90 per cent of the beef on the local market and all of Zimbabwe's export beef.

Unfortunately, the special regard in which rural communities hold their livestock has led to high cattle densities in many areas, resulting in overgrazing and consequent soil erosion and loss of fertility. The land cannot take the strain.

Clearly, correcting this deteriorating situation depends on reducing the pressure from cattle, as well as following other sound conservation and land-use practices. But this depends upon convincing people of the need to do so. Old habits die hard – values and customs shaped by long tradition are not changed overnight. Time, however, is not on Zimbabwe's side.

The Cattle Bias

Zimbabwe's preoccupation with its beef industry is wasteful and misguided, say some ecologists. They maintain that the heavy investment in time and money is going to a system of livestock production which, in the long term, is not sustainable. Low rainfall and poor soils mean that the carrying capacity in much of the country is severely limited.

Instead, they argue, the nation should be investing much more in developing its own indigenous animal resources – the wildlife which it has in abundance and which makes such efficient use of the land.

The argument, with which cattlemen tend to disagree, is a persuasive one. It is based on the fact that vastly more time, effort and capital has been invested in introducing exotic plants and animals to Africa than in nurturing its considerable indigenous resources. Africa has the richest large mammal fauna in the world – yet animal production systems focus almost entirely on cattle. It seems absurd that indigenous large animals, such as the buffalo, continue to be eliminated to make way for domestic livestock.

Two main reasons are ventured for this bias towards cattle. Firstly, wildlife was (and continues to be) viewed very largely as 'game', which carries with it connotations of sport and recreation. Secondly, the emphasis on exotic plants and animals stems from the colonial experience. Settlers brought in the animals they best knew how to handle and rear; at the same time, they were largely ignorant of the wild animals already present and of their intrinsic value. The result has been a general lack of the knowledge, expertise and technology required to manage wild animals and indigenous forests in economically productive and reliable ways.

Says Dr. David Cumming, a leading proponent of the wildlife production concept: 'It is tragic that we continue to squander our indigenous resources in order to export beef to Europe, which already has more than it can use. Given our ecological and socio-economic constraints, I do not see beef exports as a sustainable enterprise and one that should dominate our land use planning in the way it presently does.'

Of overriding importance is the *sustainability* of the system in the long term – to meet the needs and improve the lives of a growing number of people without undermining the land upon which they all depend. That may sometimes mean foregoing short-term gains, like putting beef on the tables of Europe.

Eradication of the Tsetse Fly

David Livingstone could have chosen all kinds of splendid images for the title page of his 1857 book 'Missionary Travels'. But it is an engraving of a tsetse fly which confronts the reader. Livingstone felt the tsetse was a symbol of the great constraints handicapping the development of the 'Dark Continent'.

Today, the tsetse fly *(Glossina* spp*)* still spreads the parasitic disease Trypanosomiasis. Known as sleeping sickness in people, and nagana in livestock, the disease is prevalent over an area of ten million square kilometres in tropical Africa.

Large areas of Zimbabwe's northern lowveld are infested with the flies. Traditionally, there have been few livestock here, but rising populations are putting pressure on the land – and people are settling even these infested areas. Given this pressure on land, how can the tsetse fly be eradicated? And how can the land be developed sustainably following eradication?

These questions are being tackled jointly by the Zimbabwean tsetse control and land development authorities. Only by working together will they find a long-term answer.

Tsetse control has many environmental implications. Early methods, from 1920 to 1945, relied on the destruction of wild animals. The great rinderprest outbreak of 1896, which killed large numbers of wild animals and livestock (and caused the tsetse fly to recede in many areas) had clearly demonstrated the link between the fly and vertebrate blood – its sole food and drink. As a result, as many as 800,000 wild animals were shot in this early battle against the fly, which extended along a formidable front of 700 kilometres. Later, the axe succeeded the gun, with large-scale tree felling and bush clearing and burning to displace wild animals. This deprived the fly of its essential food and also deprived it of shade and resting places.

These techniques – the only ones possible for the pioneer glossinologists – were very successful. They opened up thousands of square kilometres of former fly belt to settlement and the plough; these areas are now among the most productive commercial farmland in the country. But the techniques caused widespread environmental damage.

Then insecticidal spraying was introduced. First came Dieldrin, in 1961, which is lethally toxic to small mammals, birds and reptiles; and then, from 1968, the less costly DDT. The environmental effects of these persistent chemicals were not investigated until some years later. Then studies in Zimbabwe revealed accumulations of DDT in human milk that were twice the maximum levels recommended by the World Health Organisation. The search for an alternative to ground spraying was speeded up.

Even today, ground spraying is still carried out with DDT in selected situations between June and the end of September. If planned properly, the spraying results in total elimination.

But this backbreaking task is progressively giving way to aerial drift spraying with endosulphan. This is a more tsetse-specific chemical which breaks down quickly in the environment. It is laid at night, in minute quantities, from low-flying aircraft. Because at any one time two-thirds of a tsetse population can be underground in pupal form, five applications are put down in a carefully timed cycle.

Zimbabwe, Mozambique, Zambia and Malawi are currently undertaking an initial three-year trial, financed largely by the European Economic Community, to test the aerial spraying technique. The four countries share a major 'fly belt', which covers a total area of 322,000 square kilometres. Monitoring the environmental impact of the chemicals used is an integral part of the operation and, more recently, a study of the implications for subsequent land use has been commissioned.

Zimbabwe's own search for effective tsetse eradication, with no adverse environmental impact, has resulted in the invention of a new and safe technique – the odour-baited trap. Dr. Glyn Vale and his colleagues, in the branch of Tsetse and Trypanosomiasis Control of the Department of Veterinary Services, have developed a screen which imitates the colour and smell of cattle (a bottle of acetone mimics the animals' breath) and thereby attracts flies from hundreds of metres around. An insecticide-impregnated black screen kills the flies as they land and seek to draw blood. An extraordinarily successful experiment, covering 600 square kilometres near Chirundu, with four traps per square kilometre, resulted in a 99.9 per cent reduction in the tsetse population.

This Zimbabwean breakthrough has been described as the biggest step forward ever in tsetse control.

However, tsetse control is not an end in itself. It is but one tool of an integrated approach to land development. Following tsetse eradication, the number of possible land-use options increases. The wrong option, such as putting large numbers of cattle onto fragile land, can lead to swift overgrazing and soil erosion. Glossinologists and conservationists agree on the paramount need for carefully controlled land use by settlers and their livestock, as they move in behind the fly. Currently, the Zimbabwean Government, IUCN and the EEC are cooperating to identify ways of integrating tsetse control with land development.

It would indeed be tragic if Zimbabwe, which has done so much and laboured so long against the tsetse, should win the battle against the fly but lose the war on the ground.

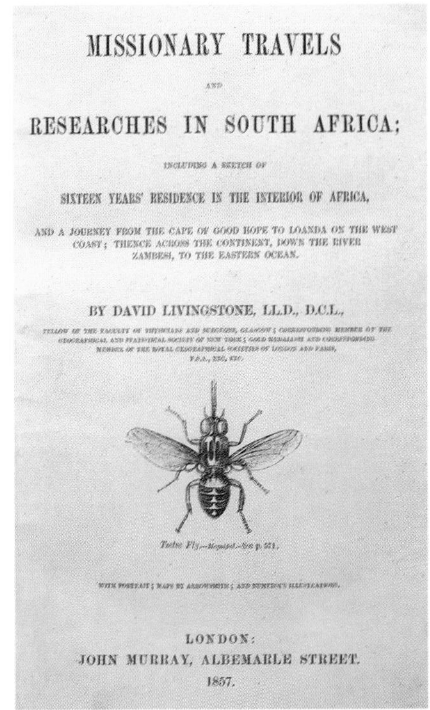

MISSIONARY TRAVELS

AND

RESEARCHES IN SOUTH AFRICA;

INCLUDING A SKETCH OF

SIXTEEN YEARS' RESIDENCE IN THE INTERIOR OF AFRICA,

AND A JOURNEY FROM THE CAPE OF GOOD HOPE TO LOANDA ON THE WEST COAST; THENCE ACROSS THE CONTINENT, DOWN THE RIVER ZAMBESI, TO THE EASTERN OCEAN.

BY DAVID LIVINGSTONE, LL.D., D.C.L.,

FELLOW OF THE FACULTY OF PHYSICIANS AND SURGEONS, GLASGOW; CORRESPONDING MEMBER OF THE GEOGRAPHICAL AND STATISTICAL SOCIETY OF NEW YORK; GOLD MEDALLIST AND CORRESPONDING MEMBER OF THE ROYAL GEOGRAPHICAL SOCIETIES OF LONDON AND PARIS, F.S.A., ETC. ETC.

Tsetse Fly.—Magnified.—See p. 571.

WITH PORTRAIT; MAPS BY ARROWSMITH; AND NUMEROUS ILLUSTRATIONS.

LONDON:
JOHN MURRAY, ALBEMARLE STREET.
1857.

Top:
David Livingstone felt that the tsetse fly was one of the biggest problems facing Africa's development. In many places, this remains the case today. (Collection of the National Archives of Zimbabwe.)

Above:
Zimbabwe's Dr. Glyn Vale has been the recipient of a major world prize (the Ciba-Geigy prize for research in animal health) for his development of environmentally-sound tsetse control methods.

Tsetse Control and Conservation

Tsetse control strategists in Zimbabwe have long been the target of bitter criticism by conservationists. They point to the destruction over the years of hundreds of thousands of wild animals, the wholesale felling of trees and clearing of large areas of bush, and the cumulative effects of heavy spraying with toxic chemicals.

Indeed, Hwange National Park, Zimbabwe's largest and one of Africa's foremost game reserves, was set aside in 1930 as a direct result of concern over the destruction of wild animals in the early tsetse control operations.

But, perhaps ironically, a key proponent for the creation of the park was Rupert Jack – first tsetse entomologist in the country. Another keen supporter of conservation is the former head of the Branch of Tsetse and Trypanosomiasis Control, Des Lovemore. Now regional co-ordinator of the EEC-assisted Regional Tsetse and Trypanosomiasis Control Programme, covering the fly belt in Zimbabwe, Zambia, Mozambique and Malawi, he was once also head of the Department of Natural Resources. He was personally involved in the creation of Chizarira National Park and the adjoining Chirisa Safari Area, in the Zambezi Valley. Both are outstanding game and research areas.

'We are probably more conscious of the environmental effects of our operations than anybody else. But we have a job of work to do. This country is in the forefront of tsetse research and control, and has much to be proud of. Sure, I have seen a lot of land hammered, but I have also seen very large areas of now highly productive farmland reclaimed from the fly. The entomologists and policymakers have always used the best means known to them, and they have been very effective. Of course, we are learning all the time.'

Like Des Lovemore, it is important for more people to be familiar with the issues on both sides. There *are* serious adverse side effects of tsetse control – but it also has many positive aspects. Conservationists, and the scientists who have led the long war against the tsetse fly, must work together to achieve sustainable development which, ultimately, depends on the success of both their causes.

Rural Development

Tea is quickly served to the visitor at the head office of ARDA, the parastatal Agricultural and Rural Development Authority. The courtesy, though welcome, is not of course unusual. What is unusual is that ARDA does make a particularly good pot of tea – with leaf grown, processed and packaged at its own Katiyo estate, in the fertile Honde Valley on the eastern border with Mozambique. The Katiyo blend goes down well in Zimbabwe – and increasingly so on the export market.

Tea is one of a dozen major crops grown by the Authority. Its 25 estates total just under one million acres and employ, at peak seasons, no fewer than 25,000 people. A tour of the ARDA estates is a tour of Zimbabwe itself. From the lush tea and coffee plantations and dairy herds of the Eastern Highlands; down to the hot and dry south-eastern lowveld, where rich clay soils under irrigation yield bumper cotton and wheat harvests; west to the sprawling ranches of Matabeleland; and up in a wide sweep back to the higher country, where better rainfall, supplemented by irrigation, permits a wide range of crops for the home and export markets.

But ARDA's mandate from the Government extends far beyond its farming activities. It is charged with no less than 'planning, coordinating, implementing, promoting and assisting agricultural and rural development in Zimbabwe'. It is a formidable directive, with implications that will have an impact on the nation in many ways and for many years to come.

The Authority's General Manager, ecologist Dr. Liberty Mhlanga, is fully aware of the scale of the task ahead. The prospects for success are exciting, the consequences of failure grim.

He accepts that the situation in the ecologically stressed areas of the country will get worse before it gets better. But he believes that the pendulum will swing back, that the mass of the people will respond to the conservation message and will adopt a more environmentally sensitive approach to save their land, 'I think Zimbabwe, in the long term, has a better chance than many countries to get to grips with its environmental problems. Although our natural resources are seriously depleted in some areas, there is still so much here to work with, to preserve. But I am afraid that the rate at which we are destroying our resources is such that by the time there is a countrywide change in attitudes and practices it may be too late in some areas.'

The phenomenon of environmental degradation which, as recently as the late 1970s, used to be seen only in communal lands has now spread into some of the commercial and marginal areas. There seems to be a general and, in some cases specific, erosion of the production base. The silting of dams and rivers, plus the adverse changes in the micro-climate, are two of the symptoms of this gradual environmental decline.

Dr. Liberty Mhlanga is keenly aware that conservation and development are interdependent, and puts this knowledge into action in his sound management of ARDA's estates. On the international scene, Dr. Mhlanga plays a role for Zimbabwe in several capacities, including that of IUCN Councillor.

ARDA's special challenges lie in the communal areas which have been degraded – in greatly varying degrees – by decades of mismanagement of generally fragile environments which have poor soils, low or erratic rainfall and increasing human and livestock populations.

ARDA works with several Government ministries and other parastatals, in overlapping efforts to arrest this deterioration and reverse the trend. Farmers are encouraged and helped to produce more, on less land, and to rest the remainder of their fields; they are encouraged to keep fewer, but healthier, livestock and to rest the remainder of their pastures. The lessons are difficult – and putting them into practice can be hard. But there is growing acceptance of them and there are some notable success stories that point encouragingly to the future.

In the Gwanda district of south Matabeleland, there is an outstanding example of an all-out effort to rehabilitate an area of about 311,000 hectares. The aim is to restore its productivity and thereby raise the living standard of its people. It is regarded as a prime test case for other districts.

The Gwanda project has involved dismantling existing village structures and bringing the people together, with facilities such as roads, water, sewage, schools, stores, community halls, a post office and a bank. Outside this fenced living complex are the arable lands. All the livestock is removed and put into fenced holding ranches, on the perimeter, to allow the land around the village to recover. The livestock is then returned after a couple of years, in better condition and reduced in number, to fenced paddocks on the replenished grazing land.

Perhaps inevitably, suspicion and resistance greeted the Gwanda project in its early stages. But now there is general acceptance of the scheme and full participation in it by the local people. They literally see its benefits growing all around them.

Tea pluckers at ARDA's lush Katiyo Estate.

ARDA also has a particular responsibility in the big resettlement exercise that is taking place in many parts of the country. Sadly, in many areas, formerly productive commercial farmland acquired by the Government for this purpose has itself suffered swift deterioration in recent years, as a result of large-scale tree felling and other forms of mismanagement. But the resettlement programme is successfully relieving environmentally depleted areas by relocating whole communities to land where they can start afresh, away from the exhausted fields and denuded pastures of their traditional homelands.

Resettlement is now taking place (or about to take place) in parts of the Zambezi Valley which have recently been cleared of tsetse fly. This harsh and hauntingly beautiful wilderness holds an important wildlife population and its ecosystem is both finely balanced and exceedingly vulnerable. There is widespread concern about the environmental damage that could result in this unique part of Zimbabwe. Dr. Mhlanga and his ARDA planners make the point that the settlement exercise in the Valley is a planned and controlled one. The carrying capacity of each area, in terms of people and their livestock, has been carefully determined. The emphasis will be on wildlife incorporation and the provision of dams for small irrigation schemes.

But misgivings persist and it remains to be seen whether people and their livestock can indeed integrate harmoniously into this essentially wild environment, or whether the impact will prove too much for the area to bear.

Chapter Five

TREES FOR THE PEOPLE

Tree Distribution

On vegetation maps, Zimbabwe appears in what is known as the Zambezian Region. Extending from three to twenty-six degrees south and from the Atlantic almost to the Indian Ocean, this region also includes the whole of Zambia and Malawi, large parts of Angola, Tanzania and Mozambique, and smaller parts of Zaire, Namibia, Botswana and South Africa.

Miombo woodland in all its forms is dominant in Zimbabwe, covering most of the central plateau higher than 1,200 metres above sea level. In deeper soils, it also occurs as low as 800 metres. Known locally as musasa-munondo woodland, it consists mainly of *Brachystegia spiciformis* and *Julbernardia globiflora*, with a discontinuous ground layer of shrubs, herbs and grasses.

As the rainfall and altitude decrease to the west and south, the musasa gives way to mufuti *(Brachystegia boehmii)* until, in the south-east lowveld, mopane woodland takes over, with *Colophospermum mopane* dominant and the massive baobab *(Adansonia digitata)* common. In the Bulawayo area in the west, *Terminalia sericea*, and various *Acacia* and *Combretum* shrubs, predominate. Teak woodlands, of *Baikiaea* species, are characteristic of the Kalahari sands of the west and south-west. In the Zambezi basin, there are extensive areas of jesse bush, which is really dry forest reduced to thickets by large mammals such as elephants.

The high-rainfall Eastern Highlands contain relict islands of evergreen montane rain forest which, although very small in relation to the rest of the country, contain more than half of the woody plant species found in Zimbabwe. These forests, fed by rainfall and mist precipitation channelled down large rock faces, include the giant mahoganies *Khaya nyasica* and *Lovoa swynnertonii*, and several other massive species such as black-bark *(Diospyros abyssinica)* and Cape mahogany *(Trichilia dregeana)*.

There are also extensive plantations of exotic pine, eucalyptus and wattle along the eastern mountains bordering with Mozambique, which form the basis of a large and well-developed forest industry.

In many parts of the country, the vegetation pattern is exceedingly complex, with multiple species associations depending on moisture, temperature, altitude and soils. In general, however, the transition from one dominant type of woodland to another is clear, and even abrupt – an unusual and, floristically, very interesting characteristic.

Deforestation

Blue woodsmoke seeping into the sunset, through the conical thatched roofs of pole-and-mud huts, is a pervading image of the Zimbabwean countryside. Inside, on small open fires under the cooking pots of peasant families, thousands of tonnes of wood are going up in smoke every day. The evening meal over, the glowing coals will then warm the sleeping occupants during the cool winter nights.

Opposite:
Forest products are becoming relatively scarce and expensive in Zimbabwe. Here, a load of pine logs is being received at the board and paper mill in Mutare.

Above:
The fire-tolerant miombo woodland is the dominant vegetation type in Zimbabwe.

Right:
More than half of all Zimbabwe's woody plant species are to be found in the isolated relicts of evergreen montane rain forest. These are situated in remote parts near the eastern border with Mozambique, such as here at Chirinda Forest.

The women of the households will have gathered the wood from up to five kilometres away – and carried it back to the village on their heads. Wood collection and building the cooking fire is a domestic ritual for the women as rhythmic, and as constant, as the seasons. Meanwhile, the village men make deeper forays into the indigenous forests, to select and chop poles to make new huts, grain bins or cattle pens. Rural life without wood is inconceivable. Indeed, it would be impossible.

Fuelwood accounts for an estimated 31 per cent of Zimbabwe's total energy consumption. It meets about 80 per cent of the energy demands of households on the communal lands. These areas make up only 42 per cent of the country's land area yet are home to nearly six out of every ten Zimbabweans.

With a rapidly growing rural population, the demand for fuelwood, and new land to grow crops, has increased considerably in recent years. In those areas where pressure on the land is heaviest, there are now severe, or even critical, fuelwood shortages. People are turning to crop

Above:
Fuelwood meets 80 per cent of the energy needs of households in communal lands. As populations continue to rise, the surrounding woodlands are becoming deforested. Foresters say that the 'sustained yield capacity' of the slow-growing woodlands is being exceeded.

Left:
Pine plantations – notably Pinus patula *from Mexico – cover vast areas of the Eastern Highlands.*

residues, and even dung, to cook their evening meals. But this only accentuates the problem, by depriving the run-down soil of organic matter and depleting it still further.

The costs associated with the national decline in wood are many and various. They include the strain on women's time and energy – which has implications for their other work in agriculture, housekeeping, productive employment and child rearing; and fuel scarcity may lead to reduced energy use, with negative social effects. There is, above all, the huge cost of environmental degradation, with its long-term impact on agricultural production and food security. This process of land degradation is inexorable – loss of trees, loss of grass, loss of soil, loss of fertility, loss of crops, loss of income; and, ultimately, loss of hope.

More than half of Zimbabwe's 55 districts are now officially fuelwood deficit areas. On a provincial level, the provinces of Manicaland, Mashonaland East, Masvingo and Midlands are categorised as not being able to meet provincial fuelwood demands. There are rural areas which still enjoy a fuelwood surplus (mainly regions of lower population density) but there is still a risk that too much land is being converted to agriculture and too much fuelwood is being sold to high-density residential suburbs, in the towns and cities, where it sells readily to lower-income households.

The problem of diminishing sources of indigenous wood was recognised as early as half a century ago, when the first woodlots of fast-growing eucalyptus were planted. The sporadic development of woodlots and nurseries continued over the years, but these forestry programmes never received the necessary level of technical support. It was not until 1982 that the Forestry Commission, the state forest authority, launched a pilot Rural Afforestation Project aimed at encouraging the establishment of woodlots by councils, schools, non-governmental agencies and individuals.

Because woodfuel will continue to be indispensable for the mass of the rural population (and for a great many urban consumers) for the foreseeable future, the improved management of existing indigenous woodlands is critical in the implementation of an integrated approach to the energy problem.

Sacred Trees

A measure of the worsening woodfuel shortage in parts of Zimbabwe is the fact that trees which once were protected from the axe, by traditional beliefs and customs, are now being chopped almost indiscriminately in some places.

Traditional restrictions on tree cutting – where particular species (or localities) were considered sacred and given complete protection – are being waived or ignored in the search for fuel. *Burkea africana, Sclerocarya birrea, Flacourtia indica,* and *Pseudolachnostylis maprouneifolia* were all considered taboo in the past but are now being felled in some areas.

On the other side of the coin, a significant and highly encouraging development is the emergence, in some areas, of strict new controls on tree cutting. These have been introduced by the communities themselves and are being enforced by a variety of local government and political agencies. In several areas, permission to fell trees must first be obtained from a local committee elected by the people themselves. The general sentiment among these groups was expressed by a villager in Gutu Communal Land: 'Villagers should hold meetings and work out their own laws with regard to the use of wood... thus they are proud of their laws and will abide by them, rather than having the laws imposed on them by others.'

Afforestation

Zimbabwe is keenly aware of the growing problem of deforestation and land degradation. A major programme, known as the Rural Afforestation Project, is tackling the source of the problem with vigour. Run in association with the World Bank, it is now making some fundamental changes in the country's approach to tree management.

The first phase of the project, which ran for three years, involved a very active schools education programme; the establishment of about 70 state nurseries, providing rural people with seedlings at subsidised prices; and the enthusiastically supported National Tree Day, with its annual planting exercises throughout the country. Above all, this phase has been enormously successful in raising public awareness of the importance of trees.

In 1989, with the project well rooted, comes a big leap to the second phase. This involves a highly significant change of emphasis on the part of the state Forestry Commission. As Deputy Director Dr. Yemi Katerere puts it: 'We have always regarded ourselves as managers of trees, looking at the

With massive afforestation as a priority, the Forestry Commission is backing the nation's efforts with a strong research programme.

tree above the ground. Now we see ourselves increasingly as managers of land, and we are telling our staff to get out of the forests and look at the trees as an integral part of the whole farming system in the communal lands.'

This fundamental swing in attitude – a wider perception of forestry in all its forms – is a major policy watershed in Zimbabwe. It is one which holds out a real hope for the future. Agro-forestry – the cultivation of trees within farming systems – is the new buzzword.

But it is only really an extension of the practices of rural families who, for a long time, have looked to the trees for fuel, food, animal fodder and other daily needs. Rural people are quite literally at home with their trees – in the communal grazing lands, around arable areas and as live fences around their huts.

The big difference, as the Rural Afforestation Project gathers momentum, is that this traditional use will be made more efficient (and more productive) with the technical backing of the Forestry Commission and the involvement of other Government agencies.

The Forestry Commission sees itself as an innovator and catalyst, rather than an indefinite implementer. The idea is for the people themselves to adopt and manage the new schemes on a financially viable basis. Nurseries, for example, will continue to be established and developed as seedling pools for community woodlots, and for other plantings, but they will later be taken over by local community administrations – as revenue-earning undertakings.

The policy swing is also reflected in the Forestry Commission's training and research programmes. Training of its foresters now includes a strong agro-forestry component and commercially driven research is giving way to a research thrust that is more social in its motivation.

Says Yemi Katerere: 'Forestry as a whole in Zimbabwe is moving away from the preoccupation of protecting trees *from* the people – to the management of the trees *by* the people and *for* the people.'

While the Forestry Commission seeks solutions for Zimbabwe, it also contributes in a vital way to afforestation research programmes in other African countries. It does this by acting as a seed bank, importing and testing seed from many sources, on behalf of countries which are bedevilled by the difficulty of getting forest seed true to type and of known geographical origin. With this facility, national afforestation efforts can often avoid mistakes on a grand scale which would take years to materialise. Countries supplied by the seed centre include all those of the SADCC region (except Angola) plus Kenya, Uganda, Ethiopia and Madagascar.

Below:
Seeds of Eucalyptus grandis *– as many as 1.25 million in a kilogram.*

Above and opposite:
30,000 hectares of eucalyptus plantation (opposite) make these Australian species some of the most widely planted trees in Zimbabwe. However, researchers feel that Zimbabwe's indigenous biological diversity has not been properly exploited. Now, trials with promising African species such as Acacia albida *(above) are revealing a range of possibilities. Farm trees of the future should be multi-purpose;* Acacia, *for example, provides good fodder and soil fertility, which eucalyptus does not, and in addition it is a better fuelwood. The main drawback of many native species – slow growth – can be reduced by selecting seed sources and by breeding programmes.*

The Search for the Perfect Tree

Without the fast-growing Australian eucalyptus, or gum tree, Zimbabwe will probably never get on top of its fuelwood shortage. Few other trees give as much wood, as quickly, as this one does.

The eucalyptus planting programme is not without opposition. Critics say that the tree's high consumption of water and nutrients is affecting the hydrological cycle and depleting soil fertility; and that, unlike many indigenous trees, its leaves and bark make poor fodder.

But planting continues on a large scale – using two species in particular: *Eucalyptus grandis*, in the better rainfall areas; and *E. camaldulensis*, in the dry zones. Zimbabwe now has about 30,000 hectares of eucalyptus plantation, including some 12,000 hectares of woodlots on the communal lands and commercial farms.

According to Lindsay Mullin, who heads the Forestry Commission research branch, there is more to the eucalyptus planting programme than many people realise. Far from damaging the local environment, he says, the tree can help to preserve it by relieving the present heavy pressure for fuelwood on indigenous woodlands. And eucalyptus trees provide more than wood. The growing honey industry in Zimbabwe, for example, depends heavily on them for pollen and nectar. Outstanding among these is *Eucalyptus melliodora*, said to be the best honey tree in the world. A single tree, such as this, that produces both good fuelwood and honey would clearly be an asset in the communal areas. Other eucalyptus species under study include those whose leaves provide essential oils that can be distilled, for medicinal and industrial purposes, and for use in perfumery.

Trials with a number of indigenous acacias are showing early promise. Foremost among these is *Acacia albida*. It thrives away from the low-lying river systems that are its natural habitat; crops can be grown right up to the tree trunk; it loses its nitrogen-rich leaves at the beginning of the summer, thus providing an excellent fertiliser; and its pods are palatable and very nutritious for livestock.

Meanwhile, the search for useful trees – both indigenous and exotic – continues with a vengeance.

Chapter Six

THE INDUSTRIAL CONNECTION

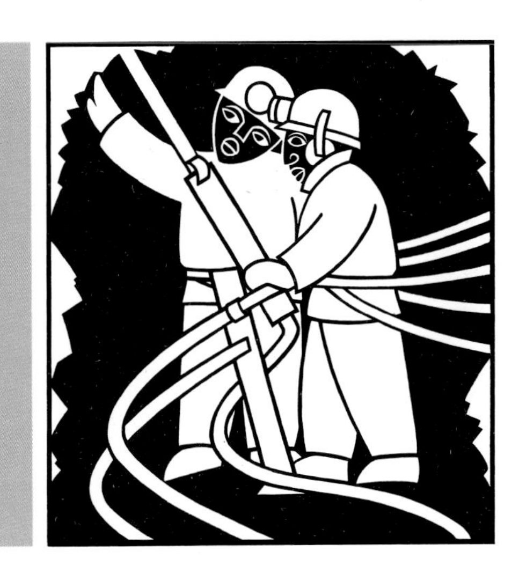

Manufacturing Industries

Every manufacturing industry depends on the input of raw materials, a very large proportion of which are natural resources. The conservation and development of these resources is therefore vitally important to the future of industry. Their yield can be optimised – and, in the case of renewable natural resources such as plants and animals, sustained.

Zimbabwe's manufacturing sector is exceptional in Africa for its diversity. It produces an impressive range of more than 6,500 products. Among them are foodstuffs, drink and tobacco, textiles (including cotton ginning), clothing and footwear, chemical and petroleum products, metals and metal products, wood and furniture, paper and paper products, non-metallic mineral products, transport and equipment.

This industrial capacity gives Zimbabwe a high degree of self-sufficiency and is also a major export earner. Manufacturing contributes the largest share of Zimbabwe's economic output and is the second largest employer, after agriculture.

Zimbabwe is no exception to the general African rule that industrial expansion in the absence of a strong agricultural sector is impossible. The connection between manufacturing and agriculture in Zimbabwe is great – and their fortunes are directly linked. Manufacturing provides a market for a significant proportion of agricultural production, while agriculture relies on essential inputs from local industry. This close interdependence is particularly significant for a country which must look to its own domestic capabilities for its wellbeing, though its efficiency and productivity is still heavily dependent upon scarce foreign exchange for imports of materials and equipment.

But Zimbabwe is fortunate in having developed that firm agricultural base. Given other essential growth factors, agriculture could drive the manufacturing and service sectors – and springboard industry to the new levels of exports necessary for overall economic growth.

The National Water Project

Clean water and proper sanitation are fundamental to human health. It is estimated that 80 per cent of sickness in the developing world is due to dirty water and poor sanitation, with at least half of all hospital beds being occupied by patients suffering from these illnesses. Most deaths occur among children under five years old, and most of these die from diarrhoeal diseases – which almost invariably are caused by a combination of contaminated drinking water and unsanitary living conditions.

Water is both a vital commodity and a powerful, destructive force. Two of Zimbabwe's most pressing priorities, perhaps ironically, are those of providing enough water for its rural people – and combating soil erosion caused by too much water on the surface.

Opposite:
As large as a small ship, the dragline at Wankie Colliery strips off thousands of tonnes of overburden to expose the coal bench.

Periodic severe droughts (sometimes for three years in a row) highlight Zimbabwe's vulnerability. Water is one of the country's scarcest resources. There are few inflowing rivers and no natural lakes; and the rainy season, which lasts from November to March, is relatively short and very unpredictable.

Increasingly, Zimbabwe depends on storage dams to meet the sharply rising demand for domestic, agricultural and industrial water. Dam construction is non-stop and there are already more than 8,000 throughout the country. Over 100 of these are classified as 'large' by the International Commission on Large Dams, of which Zimbabwe is a member. Kariba is the largest. There are also thousands of masonry weirs, just a few metres high, which are built by local communities themselves.

But dams, especially those with catchments which are subject to severe erosion, are vulnerable to siltation. On a national scale, it is not known exactly how much sediment the rivers are carrying, or what the rate of sedimentation is in the dams. But, as one hydrologist once said, 'however

Top:
Sustainable development means being able to make progress using the resources at hand, and at the same time developing the capability to avoid natural disasters. Vulnerability to drought remains a constraint to sustainable development in much of Zimbabwe. (HC)

Above:
The Blair pump is an ingenious Zimbabwean invention, requiring only simple maintenance, and allowing easy access to underground water sources.

Above right:
Despite progress, many villagers are still obliged to cart water long distances.

much it is, it is too much.'

Therefore while dam construction proceeds apace, concerted efforts are also being made to utilise underground water. The igneous rock underlying most of Zimbabwe makes underground water relatively limited, with few aquifiers to permit the operation of high-yield boreholes for major irrigation. But extensive drilling has been carried out to tap shallower reserves. More than 35,000 boreholes of low to medium yield (fitted with motor or hand pumps) and a great many wells (dug by the local people) are currently the main sources of water for most rural people.

The United Nations declared 1981-1990 as International Drinking Water Supply and Sanitation Decade. Zimbabwe, confronted by myriad problems after its newly won independence, was not slow to respond to the challenge. Said a Cabinet Minister at the time: 'It is no use sitting down under a fig tree and mourning. Nor is it enough to get down on our knees and pray. We must as men and women, and as a nation, get up and tackle the problem, using our heads and our hands. We must ourselves find the solutions.'

In response, ingenious, but simple, hand pumps and hygienic pit latrines (designed to prevent flies from getting out) were developed by the Blair Research Laboratory, in Harare. They are now making a major impact on rural health in Zimbabwe – and are being adopted in other countries too.

Then in early 1988, a national master plan for rural primary water supplies and sanitation in Zimbabwe was launched. The aim of the plan is twofold. Firstly, to provide clean water within 500 metres of the home of every rural Zimbabwean; and, secondly, to provide proper sanitation facilities for every village or community (and, where possible, every family unit) by the turn of the century. This will involve at least 50,000 more water points and 700,000 more sanitation units over and above the many thousands already in place. The cost: over a billion dollars.

There is no choice. Population growth projections, with corresponding increases in demand for food production and consequent expansion of irrigated areas, make the provision of more water – on a rapidly rising scale – one of the most urgent imperatives facing the country today.

Energy for Life

Zimbabwe's dependence on coal, and the Zambezi river, for the bulk of its energy requirements underlines the critical importance of the conservation of natural resources. The provision of energy is a key determinant of the quality of life for a country's population. Zimbabwe is fortunate. The disadvantages of not having its own exploitable petroleum resources are outweighed by the abundance of alternative energy sources. Only twelve per cent of its total energy consumption comes from imported petroleum products. No less than 88 per cent is provided by sources within the country.

Coal, from the huge reserves of the Hwange fields in the north-west, provides about 38 per cent. Most of the coal is used to generate electricity.

Wood contributes about 31 per cent, mainly for cooking and heating in the rural areas. Although it is a renewable resource, wood is being steadily depleted and Zimbabwe's natural woodlands are under severe pressure, particularly in rural areas with a high population. In some places, the indigenous trees have almost disappeared – with serious consequences.

Top:
Thirty-eight per cent of Zimbabwe's energy is supplied by the hard-working miners of Wankie Colliery.

Above:
Once Kariba Dam was completed, the Zambezi Valley was gradually inundated until an enormous lake was formed – at that time, the biggest reservoir in the world. Originally designed for hydroelectric power generation, Lake Kariba has since spawned fishing and tourism industries. But equally, unexpected problems were encountered – such as famine suffered by Tonga families resettled on poor soils. Today, the planning of large dams should include Environmental Impact Assessments to anticipate both positive and negative effects. (IB)

Above left:
Most of the coal produced at Wankie Colliery is used to generate electricity at the huge power station nearby.

Bagasse, the cane residue used as a furnace fuel on the big sugar estates in the south-east, contributes about four per cent. Ethanol (derived from sugar) and benzol (from coal) together make up two per cent. Wind power is, perhaps not surprisingly, negligible in national terms. Solar power is also poorly developed, providing a surprisingly meagre one per cent – but with Zimbabwe's sunny climate, its role could increase substantially.

However, it is to the Zambezi – already supporting two very large hydroelectric dams – that the planners are looking for almost all of the country's long-term power requirements. Hydroelectric generation is almost entirely at Kariba Dam, with power stations on both the Zimbabwean and Zambian banks. There are plans to exploit its potential by a further 6,266 MW. Under an agreement with Zambia, Zimbabwe would be entitled to half – which would more than double the country's present electricity production.

A proposal in the early 1980s to put in a third dam, at Mupata Gorge between Kariba and Cabora-Bassa, was strongly opposed by environmentalists in Zimbabwe. A dam at Mupata would flood much of the spectacular middle-Zambezi basin, including Mana Pools National Park. The proposal was eventually dropped in favour of another site, the favourite being Batoka Gorge, downriver from Victoria Falls. Mana Pools has since been declared a World Heritage Site by Unesco.

Manufacturing is the biggest energy user in the country, accounting for about 38 per cent of total consumption. Agriculture and transport account for ten per cent each, and mining nine per cent. Domestic electricity consumption is rising steadily. With the Government's large-scale electrification programme, power supplies are being extended to new growth centres in the communal lands, and to high-density urban residential districts. This is designed to alleviate hardship in rural areas and to spark economic growth. It is also bringing relief to low-income urban households

Coal is important to the national economy, but eventually its supply will be exhausted. Land, however, has the capacity for its productivity to be renewed time and time again – a capacity that needs careful nurturing through conservation. One conservation technique that is used at Wankie Colliery is land reclamation: artificially beginning a natural process that will restore productivity to land after mining.

which, without electricity, can spend as much as a third of the family budget on wood and other fuels.

The electrification programme consequently plays an important role in the overall effort to slow down the destruction of natural woodlands. But promotion of the domestic use of coal, also to save trees, is difficult. It has to be transported long distances, it requires efficient stoves and there is a strong traditional preference for wood. Researchers are currently working on the design of simple and more efficient stoves – and on the development and manufacture of alternative energy sources.

On a wider scale, Zimbabwe's five-year National Development Plan (to 1990) sets out four main objectives for the country's energy supplies: to increase electricity production from coal and hydro-power; to increase the use of coal and electricity in rural areas; to take electricity to more parts of the country; and to achieve self-sufficiency and security in energy supplies.

The Tourist Industry

'Touch the wild'. That is the compelling invitation to Zimbabwe by one tour operator. It sums up part of the attraction of the country for visitors from abroad. The wilds of Africa are only a few hours' drive from Harare International Airport.

Zimbabwe is a popular tourist destination. With about 400,000 visitors spending at least Z$100 million every year, tourism is one of the country's most important growth industries – and there is considerable potential for expansion.

Tourism in Africa is centered on natural resources, with the emphasis on wildlife. In Zimbabwe, there are some unique natural attractions. The incomparable Victoria Falls, on the Zambezi river, is the largest waterfall in the world. Kariba Dam is the site of Africa's largest man-made lake. There are the mysterious ruins of Great Zimbabwe, a variety of national parks and other protected areas, and ancient cave paintings. Accommodation is for all tastes and pockets, from five-star hotels to tented bush camps; and good roads and air services make it all accessible.

But Zimbabwe has much more to offer to anyone who would prefer to escape the rigorous (and often un-African) schedules of organised sightseeing package tours. Zimbabwe is about experiences. Backpacking through the wilderness; paddling a kayak or raft on the Zambezi river; flying through the spray of Victoria Falls; holding a baby crocodile; catching tiger-fish or trout; walking through a rain forest; riding a pony among the antelope on a game farm; seeing the sunset through the naked branches of a baobab tree; listening to the sounds of the African night; and watching elephants move deliberately through the campsite at Mana Pools.

Provided it is planned and controlled, tourism is an efficient way to exploit natural resources. It does not (or should not) deplete them. It brings in foreign currency; it provides employment in the hotel and service industries, and generally has a beneficial multiplier effect on the economy. By contributing to the preservation of the wilderness and wildlife on which it directly depends, the Zimbabwean tourist industry is a prime example in Africa of the link between resource conservation and development.

Opposite:
Most of the 400,000 visitors who arrive in Zimbabwe each year are attracted by the rich wildlife and beautiful landscapes. By deliberately concentrating on higher-cost, quality facilities – such as here at Sikumi Tree Lodge in Hwange National Park – Zimbabwe aims to earn the maximum revenue and suffer minimal environmental damage. Indeed, there is great potential for tourism revenues to pay much of the costs of conserving the national heritage.

Chapter Seven

OUR COMMON FUTURE

A Shining Example in CAMPFIRE

CAMPFIRE is the name given to the Communal Areas Management Programme for Indigenous Resources, produced by ecologists of the Department of National Parks and Wildlife Management. A flexible blue-print for improving the precarious existence of people living on remote and marginal land, and for protecting the land itself, CAMPFIRE could soon become a shining example of successful resource conservation and utilisation in Africa. It addresses the problems which arise from communal ownership of natural resources. These problems are not unique to Zimbabwe: they affect much of the African continent.

Traditional communal ownership is entirely appropriate where resources are plentiful. But the stage has been reached in the communal lands of Zimbabwe where resources are no longer plentiful. Experience shows that communities in such a situation will move towards systems of resource allocation, independently of planners, as they feel their livelihood threatened. Good planning, say the ecologists, involves recognising the inception of such a movement and assisting in a smooth transition.

CAMPFIRE is concerned with the conservation and management of wildlife, grazing, forestry and water in the more marginal communal lands. It is based on the premise that these resources belong to the community at large – and the responsibility for managing them should lie with the people living in their midst. If follows that whatever benefits arise from their custody and exploitation should accrue to the community *directly*. CAMPFIRE basically offers communities a legitimate means to obtain direct financial benefits from such planned use of their resources. The total area of land involved is some 130,000 square kilometres (about one-third of the country) and supports at least two million people.

Key to the CAMPFIRE strategy is the placing of a value on communally-owned natural resources, with an institutional structure under which communities have the incentive to manage those resources for direct sustainable returns. The grand objective is to demonstrate that conservation and development really are mutually dependent.

Communities are invited to join the CAMPFIRE. Participation is voluntary and there is considerable flexibility in the precise mechanics of the exercise, with terms resulting from negotiation rather than imposition. By developing its self-reliance, this helps to strengthen the community's participation and commitment.

Essentially, the scheme is centred on the creation of Natural Resource Cooperatives with territorial rights over defined tracts of land called Communal Resource Areas. All adults in the community become shareholders in the cooperative. They receive benefits in the form of income (shareholder dividends), employment (by the cooperative) and production (such as wildlife meat or products). Administration is through the existing Government system of Village, Ward and District Development Committees, to the provincial and national level.

CAMPFIRE still needs careful tending, but it has the potential to become a source of hope for millions of people in Zimbabwe and other parts of Africa.

Opposite:
The effects of rapid population growth are felt everywhere, from health care and food production to housing and education. Until the population can be stabilised, the provision of schools will continue to be an increasing cost.

Women and Resources

'We women carry a heavy load,' said the young woman. She could have been speaking in the literal sense. Hardly out of her teenage years, she had a baby on her back and was pregnant. She had a bundle of wood, tied together with bark, on her head. It was dusk and she was returning home to cook the evening meal. She was tired after a day of hoeing in the maize field, interrupted by household work, fetching water from the communal pump two kilometres away, and feeding the baby. The child was not well, she said. She would try to take him to the clinic tomorrow in the donkey-cart. Her husband could not help because he worked in a garage in town. But he sent her a little money, and came home by bus, when he could.

The women of the communal lands are the unsung heroines of Zimbabwe. For many years probably the poorest and most neglected section of the population, their contribution to society – in sheer hard work – has been out of all proportion to the meagre rewards and inferior status they have received in return.

This legacy of hardship and subordination is the result of both traditional social organisation and the impact of colonisation. Zimbabwean society is traditionally male-dominated and customary law made women perpetual dependents. A girl was dependent on her father and male relatives before marriage – and on her husband and his relatives after marriage. For the marriage to be legitimised, a bride price (called lobola) had to be paid; afterwards, the wife, the children and all the property belonged to the man.

It was only with motherhood that a woman could acquire some status: as the producer of sons who would continue the family line and then care for the parents in their old age. With time, she might even acquire some property. The 'cow of motherhood' was received on the marriage of a daughter; and there was also a payment for midwife services. But basically she remained dependent on her husband. Even in the event of divorce, or his death, she was entitled only to her cooking utensils, clothes and the property which she had acquired on her own.

After the British colonisation, the resulting changes imposed yet another burden on the rural woman. With her husband away from home for long periods, working in town or on a mine, she became *de facto* head of the household. This brought the added responsibility of decision-making on top of all her physically demanding work. A new kind of dependency on the man was also created, this time by his pay packet: a cash remittance, pitifully small though it was, to help her to feed the family and send the children to school.

But since independence, much has been achieved to break the customary, social, economic and legal shackles that have prevented women from fully taking part in the development of the country. The Government of Zimbabwe, through the Ministry of Community and Cooperative Development and Women's Affairs, is committed to the total emancipation of women.

But the life of the rural woman remains largely untouched by her urban sister's new spirit of freedom and opportunity. For the women of the communal lands, the daily grind of growing food in generally poor fields, maintaining the household and caring for the family, goes on with little relief.

Yet the United Nations Food and Agricultural Organisation (FAO) estimates that women produce 80 per cent of the food consumed in sub-Saharan Africa. Consequently, they are also the main users of the natural resources that support food production – and thus hold the key to conservation and development in the rural areas. It is now acknowledged in Zimbabwe that this powerful influence must not be overlooked.

At the same time, it is important to make women's productivity less arduous – and more fulfilling for them personally. That demands changes in attitude on the part of men living in the villages.

Customs and traditions are important foundations for all societies – and can be very hard to break. But there is no alternative to recognising that there can be no sustainable development in Zimbabwe if women are not full and healthy participants in the common national destiny.

Top:
Hon. Victoria Chitepo, Minister of Natural Resources and Tourism, believes that women have a specially significant role in development and the environment. She works tirelessly, both in Zimbabwe and on the international scene, to ensure that the roles of women are properly understood and improved.

Middle and above:
Women's roles are changing fast in Zimbabwe. The hard physical work traditionally performed by women in rural areas is being complemented by increasing opportunities for professional women.

Population Pressures

Zimbabwe's birth rate is one of the highest in the world. If it is not checked, the country's resources may not be able to support its population after the year 2020 and, as that time approaches, the standard of living and quality of life will decline rapidly. This is one of the most urgent warnings in Zimbabwe's National Conservation Strategy.

The average Zimbabwean mother now has 5.6 children. Consequently, the national population is increasing by just over three per cent a year – with an estimated doubling rate of only 18.7 years. Already, the total population stands at nearly nine million. Only three developing countries – Jordan, Bangladesh and Nepal – had a higher crude birth rate and general fertility rate than Zimbabwe at the time of the 1982 census.

Above:
Only three countries – Jordan, Bangladesh and Nepal – have higher birth rates than Zimbabwe's.

Above left:
Between 1969 and 1982, the population of the three biggest urban centres – Harare, Chitungwiza and Bulawayo – more than doubled.

The effects of rapid population growth are felt everywhere – in health care and social services, food production, education, housing, water supplies and sanitation, energy consumption, transport and communications, unemployment and crime. The combination of these effects is inevitably putting a heavy burden on the national economy – which is already under strain. The public and private sectors simply cannot provide sufficient jobs for the swelling numbers of school leavers and graduates.

Zimbabwe now has a vigorous and expanding family planning programme. It aims to achieve a four-child family as the norm by the turn of the century, and a two-child family by the year 2015. Meeting its family planning objectives would stabilise the population at about 23 million by the year 2075. Such a stabilised population would have a much better age distribution; at the moment, no less than 47 per cent of the population is under the age of fifteen. The provision of schools would cease to be an increasing cost; health care facilities would be able to focus on improving the quality of services; and the pressure to bring ever-increasing areas of land into cultivation would recede.

But there is another, equally important, factor in Zimbabwe's population problem – the basic distribution of its people. This is perhaps the most notable of the unfortunate legacies inherited by Zimbabwe from its colonial past. Over the years, it has had a deep significance for the use and development of the country's natural resources. Population patterns shaped in the years of land division along racial lines (a policy started in 1894 and culminated in the 1970 Land Tenure Act) persist to this day. The offensive legislation, which determined where, and on what terms, the indigenous black people could live, was repealed at independence in 1980. But its imprint – and restraining influence on development – is likely to remain for a long time to come.

The concentration of black Zimbabweans – who have never made up less than 95 per cent of the total population – on the least productive, most easily degraded land has put growing pressures on the natural resource base. Already, present land-use practices are not sustainable – and people are forced to abuse the land in order to survive in the short term.

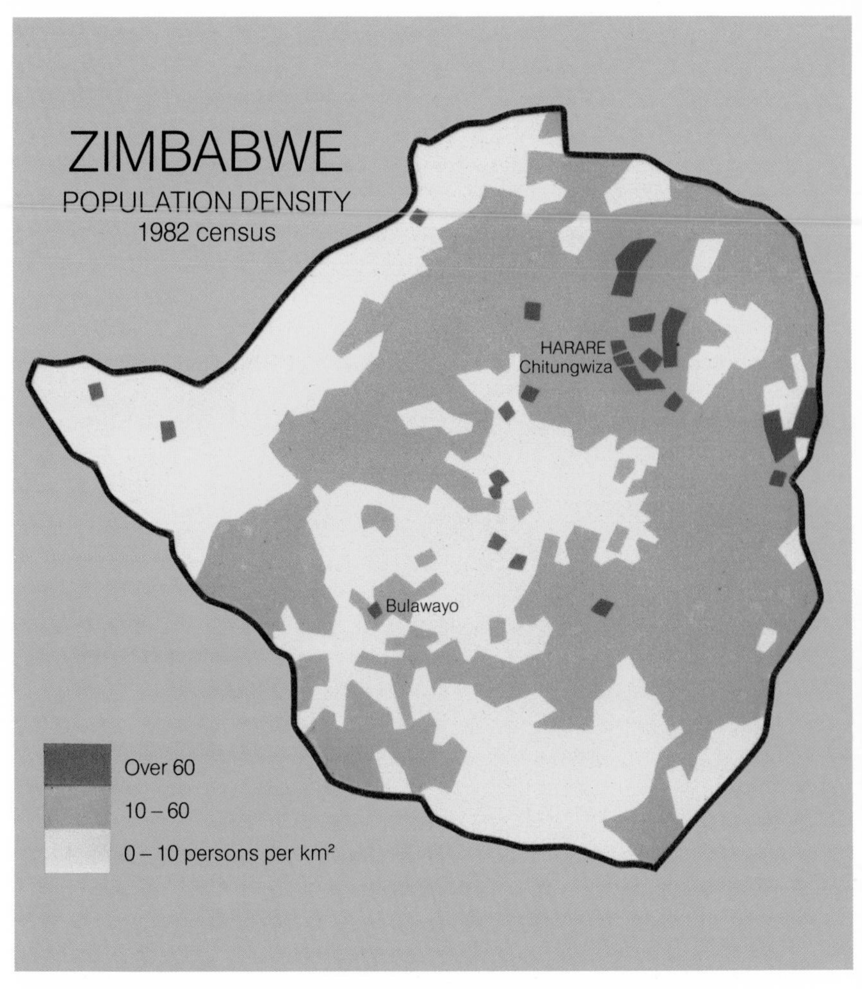

Population pressures, and diminishing opportunities, in the rural areas have resulted in a large-scale influx into the towns and cities. This is adding to the urban boom – and associated physical and social problems – caused by the displacement of people from the rural areas during the civil war years in the 1970s. Many people chose to remain in the towns and cities.

The rate of urban growth is reflected in census figures of 1969 and 1982. They put the African urban population at 0.6 million and 1.8 million respectively – an increase of 300 per cent. The figure is now much higher. In the same period, the population of the three main centres in Zimbabwe – Harare, Bulawayo and Chitungwiza – more than doubled.

The problem is being tackled on several fronts: through land-use planning and rehabilitation; agrarian reform; large-scale resettlement on former commercial farmland; environmental education and extension; and, most important, concerted promotion of family planning.

Vigorous development programmes have been designed to make the communal areas more attractive and more productive – and to contain rural-urban migration. They aim to raise living standards, to adjust population densities in relation to natural and social resources, and to decentralise economic, administrative and social activities. These efforts include the creation and servicing of remote 'growth points' in many parts of the country. Private enterprise is encouraged to participate in the development of these nuclei of commercial and manufacturing activity by providing employment, goods and services for the local communities. The growth-point concept is still in its infancy, but is already proving moderately successful and, as more centres are established and developed, holds out the promise of revitalising areas in need and improving the lives of a great many people.

The challenge faced by the Government is clearly enormous. But it is committed to the attainment of an egalitarian state, with a more equitable distribution of wealth and natural resources. As the National Conservation Strategy points out, unless there is a genuine understanding and acceptance of the need for a comprehensive population policy, the Strategy itself – and the country's aspirations for the future – will become an illusion.

The Will to Prosper

Few countries have experienced such sustained world attention as did Zimbabwe in its formative years. It was in the international spotlight for nearly two decades leading up to its true political independence in 1980. Those years were turbulent and tragic ones, the country buffeted by external economic sanctions and wracked by escalating civil warfare.

But in spite of its traumatic birth, Zimbabwe today is noted for its peace and stability – and is widely recognised as a progressive country with the will and wherewithal to prosper. It enjoys a high degree of cooperation with the international community and, since 1980, has established diplomatic relations with more than 70 countries and entered into a wide range of bilateral undertakings.

In particular, Zimbabwe plays a leading role in environmental action. This was reflected in the participation of Dr. Bernard Chidzero, Senior Minister of Finance, Economic Planning and Development, in the preparation of the World Commission on Environment and Development report called *Our Common Future*. Dr. Chidzero was one of 22 commissioners, from as many countries, who spent three years producing what has been called the most important document of our time – on the future of the world.

Our Common Future presents a global agenda for the changing of policies and practices that threaten the survival of the human race. Launching the report in Zimbabwe, in March 1988, President Robert Mugabe stressed that the country must take careful stock of its environmental situation and take 'responsible decisions to secure and sustain our resources for the present and future generations.' He said the report would be widely debated, in keeping with the Government's policy of increasing environmental awareness, through education and discussion at all levels.

Dr. Chidzero comments: '*Our Common Future* is an agenda for action on the burning issues of our time – environmental degradation, depletion of resources, pollution, the juxtaposition of boundless wealth with stark mass poverty, the debt crisis, falling commodity prices and their catastrophic effects on the toiling masses, increasing incidents of war and regional economic destabilisation. . . . These issues, brought to the fore by the report, cannot be sidetracked. We must face them squarely.'

In Zimbabwe, there are many advocates of this positive approach. It is one of the great strengths of the country's environmental movement. Local 'doom and gloom' prophets, for example, do not impress Marshall Murphree, Professor of Applied Social Sciences at the University of Zimbabwe. He says that, in particular, they do not take into account the basic strength of rural communities. These communities have an innate ability to understand the issues at stake – and to take the necessary action.

'I do not underestimate the scale of the considerable environmental problems confronting us,' says this American missionaries' son, born and raised in Zimbabwe. 'But I remain an optimist. I have seen the capacity of the people themselves to manage their environment – in sometimes very difficult conditions.' His conviction comes from the experience of years of research – from personal observation and from talking to many people.

He tells an illustrative story about a recent visit to Mhondoro Communal Land, about 170 kilometres south-west of Harare. Over the years, Marshall Murphree had been struck by the sustained excellent conservation practices of the local people. All around was clear evidence of community leadership and commitment to the management of land for the common good. He remarked on this, in complimentary terms, to a village elder. The elder's reply: 'Well, sir, we want room to dance on.'

Marshall Murphree sees embodied in that profound response Zimbabwe's hope for the future. He points to the country's ability to feed itself, to the serious attention being given to population growth, to the growing awareness of conservation issues and to the strong swing towards the recommendations of the National Conservation Strategy.

With such a positive approach, the support of the people and the necessary capacity within the country, Zimbabwe really does have the capability to manage its own environment – to create its own 'room to dance on'.

A NATIONAL CONSERVATION STRATEGY FOR ZIMBABWE

Zimbabwe's commitment to conservation of natural resources as the basis for development is spelt out in its National Conservation Strategy. Published in April 1987, with input from many people and organisations, the Strategy is no less than a blueprint for survival. It is a national recognition of the worsening problem of environmental degradation, an admission of failures which have contributed to the problem – and a statement of the resolve to try to put it right.

Zimbabwe's National Conservation Strategy was among the first to be produced in Africa. It came in response to the World Conservation Strategy (WCS), which was launched in 1980. Prepared by the International Union for Conservation of Nature and Natural Resources (IUCN), in association with the United Nations Environment Programme (UNEP) and the World Wide Fund for Nature (WWF), the WCS is based on the views of more than 700 scientists and 450 government agencies from over 100 countries. It emphasises the increasing urgency – on a global scale – to protect and use wisely the natural resources on which life depends. Specifically, it stresses the need to maintain essential ecological processes and life-support systems; to preserve genetic diversity; and to achieve the *sustainable* utilization of species and ecosystems. Many countries around the world have now prepared, or are preparing, their own strategies for the long-term conservation of natural resources.

The National Conservation Strategy for Zimbabwe echoes the message and outlines an action plan to slow down, halt and ultimately reverse the process of land damage and species extinction. Its long-term goal is to satisfy the basic needs of all the people of Zimbabwe – both present and future generations – through the conservation and wise management of natural resources. It is a formidable task but one which, says President Robert Mugabe, must be undertaken as a national and international duty. He points out that, while Zimbabwe has a proud record of environmental conservation and sound land-use systems, in over half the country the resource base is being subjected to pressures that are beyond its capabilities. 'The time for complacency is long past,' he says.

There are five key points in Zimbabwe's strategy. If the nation is to survive and prosper, it must:

- live within the ecological capacity of the land;
- recognise the value of longer-term benefits over short-term expediencies;
- examine alternative development options to optimise sustainable yield from the land;
- generate and retain high levels of technical and scientific manpower in the service of the nation; and
- provide dynamic public awareness, education and extension services.

Above all, the strategy makes it clear that Zimbabwe is one of a growing number of nations around the world whose governments view the need to tackle environmental problems as an urgent priority – not as a luxury. The support of the people is the key to putting these problems right. Success or failure therefore depends on communication, motivation – and action.

Opposite:
Zimbabwe's National Conservation Strategy aims to bring about a balance between people and livestock and the natural resources upon which they depend – soil, water, vegetation and wildlife. With careful management and the cooperation of all Zimbabweans, this balance could be reached at the highest sustainable *levels of resource productivity.*

Education is the most cost-effective long-term approach to conservation.

Acknowledgements

The information in *The Nature of Zimbabwe* has been obtained from many sources, including personal interviews, specially commissioned papers and a variety of publications – notably the National Conservation Strategy for Zimbabwe (Government of Zimbabwe, 1987) and the Tabex Encyclopedia Zimbabwe (Quest Publishing, 1987).

A great many people, all over the country, readily contributed to the project – as interviewees, writers, advisers, facilitators and local guides. We would particularly like to thank the following for their help, advice and support:

Alexandra Park Primary School; Anglo American Corporation Services Limited; Stephen Bass, IUCN Projects Manager for Southern Africa; Tom Blomefield, Tengenenge Sculpture Community; John Burton, Managing Director, and Ian Murray, Field Manager, Triangle Sugar Estates; Dr. Gordon Chavunduka, President, Zimbabwe National Traditional Healers' Association; Lydia Chikwavaire, Director, Zimbabwe Women's Bureau; Dr. Graham Child, Consultant Ecologist; The Hon. Victoria Chitepo, Minister of Natural Resources and Tourism; Meg Coates Palgrave, tree enthusiast; Dr. David Cumming, World Wide Fund for Nature Multi-species Project; Kim Damstra, biologist; Bob Drummond, Keeper, National Herbarium; Alan Elliott, Touch the Wild Safaris; Glenara Estates; Jacobus Gwerengwe, Manager, ARDA Chisumbanje Estates; Honeydew Farm; Barney Jones, Marketing Adviser, Zimbabwe Women's Bureau; Sarah Kachingwe, Secretary for Information, Posts and Telecommunications; Angeline Kamba, Director, and Gavin Douglas, Editor, National Archives of Zimbabwe; Dr. Yemi Katerere, Deputy Director, Forestry Commission; John and Pam Langley; Des Lovemore, Regional Co-ordinator, Regional Tsetse and Trypanosomiasis Control Programme; Bernard Matemera, sculptor; Dr. Francis Matipano, Director, National Museum and Monuments; Patrick Mavros, ivory carver; Dr. Liberty Mhlanga, General Manager, Agricultural and Rural Development Authority; John Mitchell, General Manager, and Peter Stranix, Manager of Personnel and Public Relations, Grindlays Bank plc; Tom Muller, Curator, and Stephen Mavi, Research Technician, National Botanic Garden and Herbarium; Lindsay Mullin, Manager, Forestry Commission Research Centre; Marshall Murphree, Professor of Applied Social Sciences, Dr. Lilian Marovatsanga, Department of Biological Sciences, and Richard Whitlow, Senior Lecturer in Geography, University of Zimbabwe; Dr. Willie Nduku, Director, and Rowan Martin, Deputy Director (Research), Department of National Parks and Wildlife Management; John Pile, Director, Zimbabwe National Conservation Trust; Dick Pitman, Editor, Zimbabwe Wildlife Magazine; Oscar Sithole, Manager, ARDA Katiyo Estates; Norman and Gill Travers, Imire Game Park; Dr. Glyn Vale, Assistant Director, Tsetse and Trypanosomiasis Control, Department of Veterinary Services; Wankie Colliery Company Limited; Bob Wannell, Chief Hydrological Engineer, Ministry of Water Resources and Development; Rose West, Exhibitions Officer, National Gallery of Zimbabwe; John White, Wildlife Producers' Association; Viv Wilson, Chipangali Wildlife Trust.

CHENGETA ZIMBABWE YAKANAKA
GCINA IZIMBABWE IINHLE

Back cover:
Epworth balancing rocks, near Harare.

Opposite:
'Keep Zimbabwe Beautiful', a wood carving in jacaranda by Gilbert Tuge of Driefontein Mission.

"The appearance of *The Nature of Zimbabwe* is most welcome. It celebrates the great natural beauty and diversity of our country; it records the achievements of our people; and it seeks to open eyes and minds to the dangers of environmental damage that grow daily around us. Above all, it emphasises the need for conservation."

Hon. Minister Victoria Chitepo, from her Foreword